Mathematics Research Developments

www.novapublishers.com

Mathematics Research Developments

J-Shaped Distributions and Their Applications
Mohammad Ahsanullah, PhD and Mohammad Shakil
2023. ISBN: 979-8-89113-013-5 (Softcover)
2023. ISBN: 979-8-89113-182-8 (eBook)

Contemporary Algorithms: Theory and Applications. Volume III
Christopher I. Argyros, Samundra Regmi,
Ioannis K. Argyros, PhD and Santhosh George, PhD
2023. ISBN: 979-8-88697-906-0 (Hardcover)
2023. ISBN: 979-8-88697-985-5 (eBook)

Four-Dimensional Atomic Structure and Law
Kunming Xu, PhD
2023. ISBN: 979-8-88697-989-3 (Hardcover)
2023 ISBN: 979-8-89113-062-3 (eBook)

Stochastic Processes: Fundamentals and Emerging Applications
Mikhail Moklyachuk, DSci. (Editor)
2023. ISBN: 978-1-68507-982-6 (Hardcover)
2023. ISBN: 979-8-88697-475-1 (eBook)

Future Relativity, Gravitation, Cosmology
Valeriy V. Dvoeglazov, PhD, M. de G. Caldera Cabral,
J. A. Cázares Montes, and J. L. Quintanar González (Editors)
2023. ISBN: 979-8-88697-455-3 (Hardcover)
2023. ISBN: 979-8-88697-527-7 (eBook)

More information about this series can be found at
https://novapublishers.com/product-category/series/mathematics-research-developments/

Mohammad Ahsanullah
and Mohammad Shakil

J-Shaped Distributions and Their Applications

www.novapublishers.com

DOI: https://doi.org/10.52305/KILR7482

NOTICE TO THE READER

Library of Congress Cataloging-in-Publication Data

ISBN: 979-8-89113-013-5

Published by Nova Science Publishers, Inc. † New York

The authors dedicate this book to their wives, children and late parents

Contents

Preface

In many fields of research, such as, biology, computer science, control theory, economics, engineering, genetics, hydrology, medicine, number theory, statistics, physics, psychology, reliability, risk management, etc., the shapes of probability distributions of non-normal data exhibit J-shaped distributions. The shapes of such distributions may be skewed to the left or the right depending on whether large percentage of data is at the lower or upper extreme. In this hand book, we have studied the J-shaped distributions and their applications. As a motivation, we have discussed several real-world examples which can be modeled through J-shaped distribution. We have presented the mathematical formulation of the family of J-shaped probability distributions which was first proposed by Topp and Leone (1955). We also have discussed several variations of Topp–Leone's family of J-shaped distribution. We have considered the general form of J-shaped distribution and derived its moments independently. We also have discussed other distributional properties of the J-shaped distribution. Some distributional properties of order statistics of the J-shaped distribution such as moment, variance, product moments, and covariance are also provided. To describe the shapes of the J-shaped distribution, the plots of the *cdf* and *pdf* for various values of the parameter have been provided. Entropy provides an excellent tool to quantify the amount of information (or uncertainty) contained in a random observation regarding its parent distribution (population). A large value of entropy implies greater uncertainty in the data. As such, Shannon entropy of the J-shaped distribution is provided. The distributional properties of order statistics of the J-shaped distribution such as moment, variance, product moments, and covariance, have also been presented. The numerical computations of these for selected values of the parameters are provided. The distributional properties of the record values of the J-shaped distribution are also investigated. Some discussions on the sum, product and ratio of the J-shaped distributions are provided. Characterizations of the J-shaped distribution are given by using the method of truncated moment, order statistics and record values, where we have considered a product of reverse hazard rate and another function of the truncated point. We have also derived the J-shaped distribution as a solution of the generalized Pearson system of differential equation. Finally, the percentile points of the J-shaped distribution are provided.

The organization of the chapters is following Chapter 1 contains the introduction and motivation, and literature review. Chapter 2 contains a mathematical formulation of the family of J-shaped probability distributions, including some variations of the J-shaped distribution, considered by various authors. In chapter 3, we discuss some distributional properties of the J-shaped distribution. In chapter 4, we discuss order statistics of the J-shaped distribution. In chapter 5, we discuss record values of the J-shaped distribution. Chapter 6 contains a brief discussion of the sum, product and ratio of the J-shaped distributions. In chapter 7, we present characterizations of the J-shaped distribution. Chapter 8 contains the computations of the percentage points for the J-shaped distribution. Some concluding remarks are given in chapter 9.

We hope that the findings of this book will be useful for the practitioners in various fields of studies and further enhancement of research in the fields of probability, statistics, and their applications.

Mohammad Ahsanullah, PhD, Professor Emeritus
Mohammad Shakil, PhD, Professor

May 2023

Acknowledgments

The authors would like to express their gratitude to their respective institutions, Rider University and Miami Dade College, respectively, for all support and assistance received throughout the preparation and completion of this book.

Chapter 1

Introduction

1. Introduction

In many fields of research, such as biology, computer science, control theory, economics, engineering, genetics, hydrology, medicine, number theory, statistics, physics, psychology, reliability, risk management, etc., the shapes of probability distributions of non-normal data exhibit J-shaped distributions. The J-shaped distribution is also known as Topp–Leone distribution, since it was first proposed by Topp and Leon (1955). The shapes of such distributions may be skewed to the left or the right depending on whether large percentage of data is at the lower or upper extreme.

The J-shaped distribution is defined as follows: "A frequency distribution that is extremely asymmetrical in that the initial (or final) frequency group contains the highest frequency, with succeeding frequencies becoming smaller (or larger) elsewhere; the shape of the curve roughly approximates the letter "J" lying on its side; McGraw-Hill Dictionary of Scientific & Technical Terms (2003)." Also, see Yule & Kendall (1950) (Chapter 4 on "Frequency Distributions"), and Ghosh (2005), among others. According to Bluman (2023, Page 65), "A *J-shaped distribution* has a few data values on the left side and increases as one moves to the right. A *reverse J-shaped distribution* is the opposite of the J-shaped distribution."

1.1. Some Real-Life Examples

Examples of skewed to the left and skewed to the right are shown in Figures 1.1.1 and 1,1,2 below;

Example 1.1.1. In Figure 1.1.1. below, we have profile of daily scores (n = 701). The number of babies with each score is shown at the top of each column. Most babies are completely healthy (implied by a score of 1) and this is shown by the majority of measurements at the lower extreme.

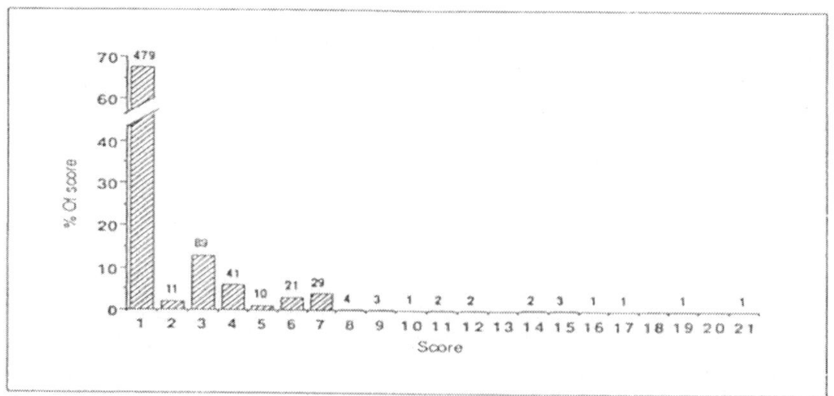

Source: Thornton, A., Morley, C., Green, S., Cole, T., Walker, K., and Bonnett, J., Field trials of the Baby Check score card: mothers scoring their babies at home, Archives of Disease in Childhood, 1991; 66: 106-110.

Figure 1.1.1. Profile of daily scores (n = 701).

Example 1.1.2. In Figure 1.1.2 below, we have linear analogue scores for pain relief after Wilson's Osteotomy. Most individuals had 100% (full) pain relief after the operation.

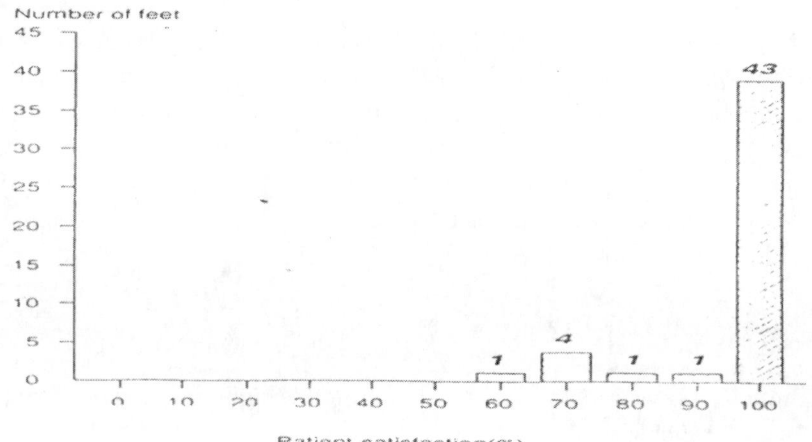

Source: Tibrewal, S., and Foss, M., Day care surgery for the correction of hallux valgus, Health Trends, 1991; Vol 23 No. 3: 117-119.

Figure 1.1.2. Linear analogue scores for pain relief after Wilson's Osteotomy.

1.2. Literature Review

Many authors have studied the J-shaped (or Topp–Leone) distribution and its properties. As pointed out above, the J-shaped distribution was first proposed by Topp and Leon (1955). It appears from literature that not much attention was paid since then, until Nadarajah and Kotz (2003) studied the distributional properties of the J-shaped distribution. Further development continued with the contributions of many authors, among them Kotz and Van Dorp (2004), Ghitany et al. (2005), Kotz and Nadarajah (2006), Zhou et al. (2006), Kotz and Seier (2007), Nadarajah (2009), Zghoul (2010, 2011), Genç (2012), Giles (2012), and Sindhu et al. (2013), are notable. For estimation of parameters of the J-shaped distribution, the interested readers are referred to Nadarajah and Kotz (2003) and Giles (2012). For moments, see Nadarajah and Kotz (2003). For the goodness-of-fit for the Topp-Leone Distribution with Unknown Parameters, see Al-Zahrani (2012). For the behavior of kurtosis, see Kotz and Seier (2007). For Bayesian estimates under trimmed samples, one can visit Sindhu et al. (2013). For some reliability measures and their stochastic orderings, we refer to Ghitany et al. (2005). Order statistics based on the J-shaped distribution can be found in Zghoul (2010), whereas, for the moments of order statistics of the J-shaped distribution, one can visit Genç (2012). For record values from a family of the J-shaped distribution, the interested readers are referred to Zghoul (2011). For moments of kth lower record values from the J-shaped distribution, see Kumar (2014). For stress-strength modeling, see Genc (2013). The sums, products and ratios of the independent random variables X and Y from the J-shaped distribution are provided in Zhou et al. (2006).

1.3. Remark

J-shaped distributions cannot be transformed to normality. This is because of the fact that, since transformations never change the order of the sample, any transformation of a J-shaped distribution will still be J-shaped. The extreme measurements will all transform to the same new value and will always be at one extreme of the transformed sample (https://epilab.ich.ucl.ac.uk/course material/statistics/index.html).

1.4. Conclusion

In this chapter, we have introduced the J-shaped distributions. As a motivation, we have discussed several real-world examples which can be modeled through J-shaped distribution. We believe that the findings of this chapter would be very useful for the practitioners in various fields of studies and further enhancement of research in distribution theory, and its applications.

Chapter 2

Mathematical Formulations

2. Introduction

In this chapter, we provide mathematical formulations of the family of J-shaped probability distributions. Some variations of the J-shaped distribution, considered by various authors, are also given. To describe the shapes of such variations of the J-shaped distributions, the plots of the related cdf's and pdf's for various values of the parameter have been provided, from which one can easily see the effects of the parameters such as the J-shapes of the pdf's of different variations of the J-shaped distributions. We have also provided the failure rate function of the J-shaped distribution.

2.1. Topp-Leone's J-Shaped Distribution

A mathematical formulation of the family of J-shaped probability distributions was first proposed by Topp and Leone (1955). They also derived its first four integer-order moments and showed its suitability to model failure data. In what follows, we first define Topp-Leone's J-shaped distribution. For an absolutely continuous random variable X, the family of cumulative distribution functions (cdf's) considered by Topp and Leone (1955), and exhibiting J-shaped distributions, is defined as follows:

$$F(x) = \begin{cases} 0, & x < 0, \\ \left[\dfrac{\alpha x}{\beta}\left(2 - \dfrac{x}{\beta}\right)\right]^{\delta} + (1-\alpha)\dfrac{x}{\beta}, & 0 \le x \le \beta < \infty, \\ 1, & x > \beta, \end{cases} \tag{2.1.1}$$

where $0 < \delta \le 1$ and $0 < \alpha \le 1$. The corresponding family of probability density functions (pdf 's) obtained by differentiation of Eq. (2.1.1) is given by

$$f(x) = \frac{2\alpha\delta}{\beta}\left(1 - \frac{x}{\beta}\right)\left(\frac{x}{\beta}\left(2 - \frac{x}{\beta}\right)\right)^{\delta - 1} + \frac{(1-\alpha)}{\beta}. \backslash \tag{2.1.2}$$

If an absolutely continuous random variable X has the Topp-Leone's J-shaped distribution with the pdf as defined above in Equation (2.1.2) with parameters $0 \le x \le \beta < \infty$, $0 < \delta \le 1$, and $0 < \alpha \le 1$, we will denote it by $X \sim TL(\delta, \alpha, \beta)$. Using the Maple software, the shape of the pdf $f(x)$ in Eq. (2.1.2).

2.1.1. Some Remarks on Topp-Leone's J-Shaped Distribution

A. By differentiating Equation (2.1.2) successively twice, we obtain the following functions:

$$f^{/}(x) = \frac{2\alpha\delta}{\beta^2}\left(\frac{x}{\beta}\left(2 - \frac{x}{\beta}\right)\right)^{\delta - 2}\left(\left(1 - \frac{x}{\beta}\right)^2 (2\delta - 1) - 1\right), \tag{2.1.3}$$

and

$$f^{//}(x) = \frac{4\alpha\delta}{\beta^3}(\delta - 1)\left(1 - \frac{x}{\beta}\right)\left(\frac{x}{\beta}\left(2 - \frac{x}{\beta}\right)\right)^{\delta - 3}\left(\left(1 - \frac{x}{\beta}\right)^2 (2\delta - 1) - 3\right). \tag{2.1.4}$$

As pointed out by Topp and Leone (1955), from the above equations, we observe the following properties of the J-shaped distribution:

(i) From Eq. (2.1.2), it is seen that $f(x)$ is positive for all x such that $0 < x < \beta$.

(ii) From Eq. (2.1.3), $f'(x)$ is seen to be negative for all x such that $0 < x < \beta$.

(iii) From Eq. (2.1.4), $f''(x)$ is seen to be positive for all x such that $0 < x < \beta$.

Thus, all probability density functions (that is, frequency curves) of the family (2.1.2) have graphs which are positive between 0 and β, have negative slopes between 0 and β, and for each of them the slope is al- ways increasing between 0 and β. These curves thus may be called J-shaped.

B. Furthermore, for a discussion on the "hazard rate function" of the J-shaped distribution and its properties, such as its "bathtub" shape, concavity, increasing-decreasing behavior, minimum value, etc., the interested readers are referred to Nadarajah and Kotz (2003), among others.

C. From Eq. (2.1.2), however, it can be seen that the probability density functions (that is, frequency curves) of the family (2.1.2) do not drop to zero at the right extreme of range except in the case $\alpha = 1$. Some curves of this type, having a positive ordinate at the right extreme of range, are called U-shaped by Pearson, for which the interested readers are referred to Johnson, et al. (1994, 1995).

D. Since a mathematical formulation of the family of J-shaped probability distributions was first proposed by Topp and Leone (1955), a random variable X having a J-shaped distribution with its distribution function and probability density function given by (2.2.1) and (2.2.2.), respectively, is also known in literature as Topp–Leone distribution.

2.2. General Form of J-Shaped Distribution

An absolutely continuous random variable X is said to have the general form of the J-shaped distribution if its distribution and probability density functions are, respectively, given by

$$F(x)=\begin{cases}0, & x<0,\\ \left[\frac{x}{\beta}\left(2-\frac{x}{\beta}\right)\right]^{\delta}, & 0\le x\le\beta<\infty, 0<\delta\le 1,\\ 1, & x>\beta,\end{cases} \tag{2.2.1}$$

and

$$f(x)=\frac{2\delta}{\beta}\left(1-\frac{x}{\beta}\right)\left(\frac{x}{\beta}\left(2-\frac{x}{\beta}\right)\right)^{\delta-1}, \tag{2.2.2}$$

which is obtained by taking $\alpha=1$ in the Equation (2.1.3) for the *pdf* of Topp-Leone's J-Shaped distribution. The distribution is called J-shaped because $f(x)>0$, $f^{/}(x)<0$, and $f^{//}(x)>0$, for all $0<x<\beta$, where $f^{/}(x)$ and $f^{//}(x)$ denote the first and second derivatives of the *pdf* $f(x)$ respectively.

2.3. As a Solution of the Generalized Pearson Differential Equation

It can be easily seen that the probability density function (2.2.2) of the general form of the J-shaped distribution satisfies the following generalized Pearson differential equation:

$$\frac{f^{/}(x)}{f(x)}=\frac{a_0+a_1 x+a_2 x^2}{b_0+b_1 x+b_2 x^2+b_3 x^3},$$

where

$$a_0=2\beta^2(\delta-1), a_1=2\beta(1-2\delta), a_2=2\delta-1,$$
$$b_0=0, b_1=2\beta^2, b_2=-3\beta, b_3=1.$$

2.4. Some Variations of Topp–Leone's J-Shaped Distribution

In what follows, we will present some variations of Topp–Leone's J-shaped distribution which have been considered by various authors.

(i) Topp and Leon (1955):

By taking $\beta = 1$ in Equations (2.1.2) and (2.1.1.), the *cdf* and *pdf* of the J-shaped distribution are respectively given by

$$F(x) = [\alpha x(2-x)]^{\delta} + (1-\alpha)\, x, \quad 0 \le x \le 1\ , \tag{2.4.1}$$

and

$$f(x) = 2\alpha\,\delta\,(1-x)\,(x(2-x))^{\delta-1} + (1-\alpha), \tag{2.4.2}$$

where $0 < \delta \le 1$ and $0 < \alpha \le 1$.

To describe the shapes of the J-shaped distribution as provided in Eq. (2.4.1) above, the plots of the (2.4.1) and *pdf* (2.1.) for some values of the parameters α and δ are provided in Figure 2.4.1. The effects of the parameters such as the J-shape of cdf (2.4.1) are easily seen from these graphs. Similar plots can be drawn for others values of the parameters.

(ii) Nadarajah and Kotz (2003); Ghitany et al. (2005); and Zhou et al. (2006):

By taking $\alpha = 1$ and $\beta = 1$ in Equations (2.1.2) and (2.1.1), the *cdf* and *pdf* of the J-shaped distribution are respectively given by

$$F(x) = [x(2-x)]^{\delta}, \quad 0 \le x \le 1\ , \tag{2.4.3}$$

and

$$f(x) = 2\delta\,(1-x)\,(x(2-x))^{\delta-1}, \tag{2.4.4}$$

where $0 < \delta < 1$. To as describe the shape J-shaped distribution as provided in Eq. (2.4.4) above, the plots of the (2.4.1) and (2.4.4) for various values of the parameter δ are provided in Figures 2.4.1, 2.4.2, 2.4.3 and 2.4.4 respectively. The effects of the parameters such as the J-shape of the (2.4.1) and (2.4.4) can easily be seen from these graphs. Similar plots can be drawn for other values of the parameters.

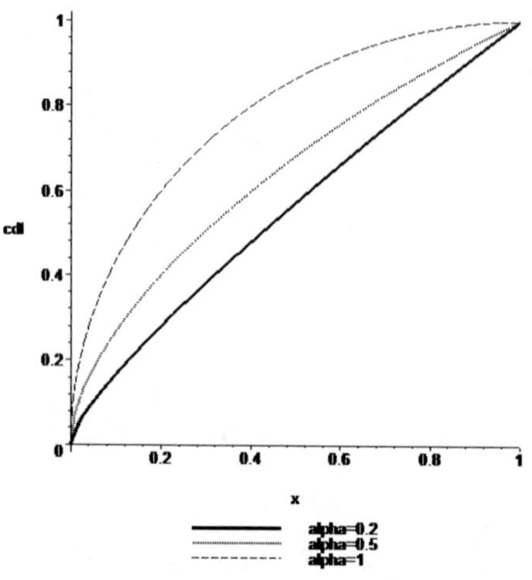

Figure 2.4.1. Plots of the J-Shaped Distribution's *cdf* (2.4.3) for $\alpha = 0.2, 0.5, 1$, and $\delta = 0.5$.

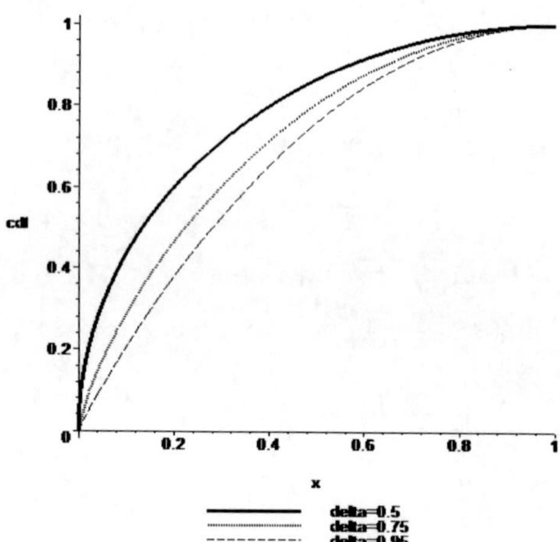

Figure 2.4.2. . Plots of the J-Shaped Distribution's *cdf* (2.4.3) for $\delta =$ $0.5, 0.75$ *and* 0.95

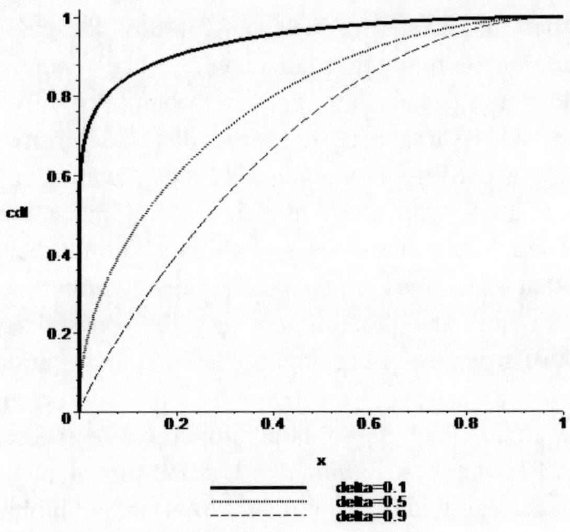

Figure 2.4.3. Plots of the J-Shaped Distribution's cdf (2.4.3) for $\delta = 0.1, 0.5, 0.9$.

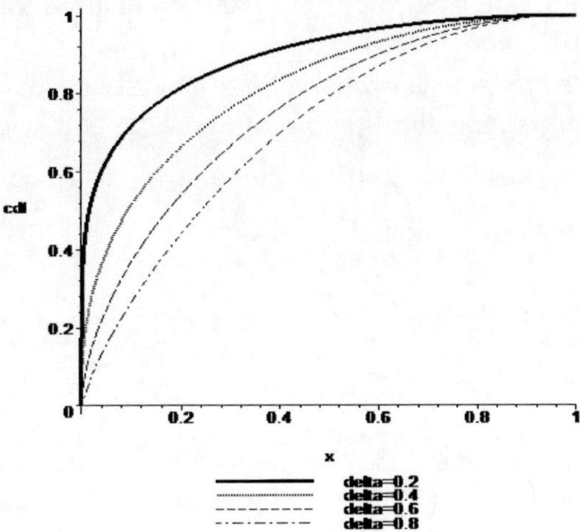

Figure 2.4.4. Plots of the J-Shaped Distribution's cdf (2.4.3) for $\delta = 0.2, 0.4, 0.6, 0.8$.

2.5. Failure Rate Function

As pointed out by Nadarajah (2009), "The failure rate function is an important quantity characterizing life phenomena. It can be loosely interpreted as the conditional probability of failure, given survival to the time x. Ideally one would like a "bathtub" shape for h(x) that captures the three distinct hazard regimes: the region of infant mortality (where h(x) decreases with x), the random failure region (where h(x) does not change rapidly with x) and the wear-out region (where h(x) increases with x due to deterioration processes)."

Furthermore, according to Nadarajah (2009), "Almost all of the standard distributions in statistics do not exhibit a bathtub shape for h(x). Even the traditional Weibull distribution does not exhibit a bathtub shape for h(x). Thus, it is important that one knows which distributions exhibit this shape. This will be important for both the practitioners and the theoreticians. The aim of Nadarajah (2009)'s paper is to review the known distributions that exhibit a bathtub shape for h(x). Apart from helping to provide better modeling, it is hoped that this review will serve as an important reference and encourage developments of further distributions with a bathtub shape for h(x)." One of such distributions identified by Nadarajah (2009) that exhibits a bathtub shape for h(x) is the J-shaped distribution.

Motivated by the above remarks of Nadarajah (2009), in what follows, we provide the failure rate function of the J-shaped distribution. For details, see Nadarajah (2009), among others.

Failure Rate Function (Definition): For an absolutely continuous random variable X, representing the lifetime of some system with the probability density function (*pdf*) $f(x)$ and the cumulative distribution function (*cdf*) $F(x)$. The failure rate function of X is defined by

$$h(x)=\frac{f(x)}{1-F(x)}. \tag{2.5.1}$$

As stated in Section 2.3, Equations (2.5) and (2.6), an absolutely continuous random variable X is said to have the general form of the J-shaped distribution if its distribution and probability density functions are, respectively, given by

$$F(x)=\begin{cases}0, & x<0,\\ \left[\frac{x}{\beta}\left(2-\frac{x}{\beta}\right)\right]^{\delta}, & 0\le x\le\beta<\infty, 0<\delta\le 1,\\ 1, & x>\beta,\end{cases} \tag{2.5.2}$$

and

$$f(x)=\frac{2\delta}{\beta}\left(1-\frac{x}{\beta}\right)\left(\frac{x}{\beta}\left(2-\frac{x}{\beta}\right)\right)^{\delta-1}. \tag{2.5.3}$$

Now substituting (2.5.2) and (2.5.3) in Equation (2.5.1), and simplifying, the failure rate function of the J-shaped distribution is obtained as follows (see Nadarajah, 2009):

$$h(x)=\left(\frac{2\delta}{\beta}\right)\frac{y\left(1-y^2\right)^{\delta-1}}{1-\left(1-y^2\right)^{\delta}}, \tag{2.5.4}$$

where $y=1-\frac{x}{\beta}$.

Remarks: As pointed out by Nadarajah (2009), it is observed that the failure rate function, $h(x)$, as given by 2.5.4), of the J-shaped distribution, has the following properties:

The failure rate function, $h(x)$, exhibits a bathtub shape for all $\delta\in(0,1)$;

In fact, $h(x)\to\infty$, as $x\to 0$, for all $\delta\in(0,1)$;

$h(x)\to\infty$, as $x\to\beta$, for all $\delta\in(0,1)$; and,

$h(x)$ attains a minimum at $x=x_0$, where $y_0=1-\frac{x_0}{\beta}$ is the root of the equation $(1-y)^{\delta}=1-\frac{2\delta y}{(1+y)}$.

2.6. Conclusion

In this chapter, we have presented the mathematical formulation of the family of J-shaped probability distributions, which was first proposed by Topp and Leone (1955). We also have discussed several variations of Topp–Leone's family of J-shaped distributions. To describe the shapes of the J-shaped distributions, the plots of the cdf and pdf for various values of the parameters involved are provided. The effects of the parameters such as the J-shape of the pdf can easily be seen from these graphs. Motivated by the importance of the failure rate function in characterizing the life phenomena, we have also provided the failure rate function of the J-shaped distribution. To illustrate the bathtub shapes of the failure rate function of the J-shaped distribution, its plots are also provided for different values of the parameters. We believe that the findings of this chapter would be very useful for the practitioners in various fields of studies and further enhancement of research in distribution theory, and its applications.

Chapter 3

Distributional Properties

3. Introduction

As discussed in chapter 2, Topp and Leone (1955) were the first to propose the family of J-shaped probability distributions which is defined as follows:

For an absolutely continuous random variable X, the family of cumulative distribution functions (cdf 's) considered by Topp and Leone (1955), and exhibiting J-shaped distributions, is defined as follows:

$$F(x) = \begin{cases} 0, & x < 0, \\ \left[\frac{\alpha x}{\beta}\left(2 - \frac{x}{\beta}\right)\right]^{\delta} + (1-\alpha)\frac{x}{\beta}, & 0 \le x \le \beta < \infty, \\ 1, & x > \beta, \end{cases} \tag{3.0.1}$$

where $0 < \delta \le 1$ and $0 < \alpha \le 1$. The corresponding family of probability density functions (pdf 's) obtained by differentiation of Equation (3.0.1.) is given by

$$f(x) = \frac{2\alpha\delta}{\beta}\left(1 - \frac{x}{\beta}\right)\left(\frac{x}{\beta}\left(2 - \frac{x}{\beta}\right)\right)^{\delta - 1} + \frac{(1-\alpha)}{\beta}. \tag{3.0.2}$$

Many authors have discussed the derivations of the moments of the J-shaped distribution, given by the Equations (3.0.1) and (3.0.2.)). For example, Topp and Leone (1955) derived its first four integer-order moments, that is, $E\left(X^n\right)$,. Nadarajah and Kotz (2003) derived explicit algebraic expressions for $E\left(X^n\right)$, when $n \geq 1$ is any integer, and, in particular, also derived explicit algebraic expressions for $E\left(X^n\right)$, when $n = 1, 2, \ldots, 10$, for $E\left(\{X - E(X)\}^n\right)$, when $n = 2, 3, 4$, and provided an analysis of the skewness and kurtosis measures of the J-shaped distribution. For details, the interested readers are referred to Topp and Leone (1955), and Nadarajah and Kotz (2003), among others.

In what follows, without loss of generality, we will consider the general form of the J-shaped distribution, as stated in chapter 2, and derive its moments independently. See also, for example, Topp and Leone (1955), and Nadarajah and Kotz (2003), among others.

General Form of J-shaped distribution (Definition):

An absolutely continuous random variable X is said to have the general form of J-shaped distribution if its cdf and pdf are, respectively, given by

$$F(x) = \left[\frac{x}{\beta}\left(2 - \frac{x}{\beta}\right)\right]^{\delta}, \tag{3.0.3}$$

$$f(x) = \frac{2\delta}{\beta}\left(1 - \frac{x}{\beta}\right)\left(\frac{x}{\beta}\left(2 - \frac{x}{\beta}\right)\right)^{\delta - 1}, \quad 0 \leq x \leq \beta < \infty, 0 < \delta \leq 1 \tag{3.0.4}$$

3.1. Moments

For any real number, $s > 0$, using the above Equation (3.0.3). the moment, $E\left(X^s\right)$, of order s of the random variable X having the general form of J-shaped distribution, is derived as follows:

$$E\left(X^s\right) = \int_0^{\beta} x^s f(x)\,dx$$

$$= \int_0^{\beta} x^s \left[\frac{2\delta}{\beta}\left(1 - \frac{x}{\beta}\right)\left(\frac{x}{\beta}\left(2 - \frac{x}{\beta}\right)\right)^{\delta - 1}\right] dx$$

$$= \frac{2\delta\,\beta^{s}}{\beta}\int_{0}^{\beta}\left(\frac{x}{\beta}\right)^{s+\delta-1}\left(1-\frac{x}{\beta}\right)\left(\frac{x}{\beta}\left(2-\frac{x}{\beta}\right)\right)^{\delta-1}dx\,. \qquad (3.1.1)$$

Substituting $\frac{x}{\beta} = u$ in Equation (3.1.1), and simplifying, we obtain

$$E\left(X^{s}\right) = 2\delta\,\beta^{s}\int_{0}^{1}u^{s+\delta-1}(1-u)\,(2-u)^{\delta-1}du$$

$$= 2\delta\,\beta^{s}\left[\int_{0}^{1}u^{s+\delta-1}(2-u)^{\delta-1}du - \int_{0}^{1}u^{(s+\delta+1)-1}(2-u)^{\delta-1}du\right]$$

$$= 2^{\delta}\,\delta\,\beta^{s}\left[\int_{0}^{1}u^{s+\delta-1}\left(1-\frac{u}{2}\right)^{\delta-1}du - \int_{0}^{1}u^{(s+\delta+1)-1}\left(1-\frac{u}{2}\right)^{\delta-1}du\right]. \qquad (3.1.2)$$

Now, substituting $\frac{u}{2} = t$ in Equation (3.1.2) and simplifying, we obtain

$$E\left(X^{s}\right)$$

$$= 2^{s+2\delta}\,\delta\,\beta^{s}\left[\int_{0}^{1/2}t^{s+\delta-1}(1-t)^{\delta-1}dt - 2\int_{0}^{1/2}t^{(s+\delta+1)-1}(1-t)^{\delta-1}dt\right]. \qquad (3.1.3)$$

By using the definition of the incomplete beta function, $B_{x}(a, b)$, which is defined as follows

$$B_{x}(a, b) = \int_{0}^{x}z^{a-1}(1-z)^{b-1}dz = \frac{x^{a}}{a}\,{}_{2}F_{1}(a, 1-b; a+1; x),$$

where ${}_{2}F_{1}(a, 1-b; a+1; x)$ denotes the Gauss hypergeometric function, see, for example, Abramowitz and Stegun (6.6.1/P. 263, 15.1/P. 556, 1972), Gradshteyn and Ryzhik (8.391/P. 950, 9.1/P. 1039, 2000), among others, we obtain, from Equation (3.1.3.), the following expression for the moment, $E\left(X^{s}\right)$, in terms of the incomplete beta function, $B_{x}(a, b)$:

$$E\left(X^{s}\right)=2^{s+2\delta}\,\delta\,\beta^{s}\left[B_{1/2}\left(s+\delta,\,\delta\right)-2\,B_{1/2}\left(s+\delta+1,\,\delta\right)\right], \tag{3.1.4}$$

which, in view of the following convergent power series expansion of the incomplete beta function

$$B_{z}\left(a,b\right)=B\left(a,b\right)-\frac{z^{a}\left(1-z\right)^{b}}{b}\sum_{j=0}^{\infty}\frac{\left(a+b\right)_{j}}{\left(b+1\right)_{j}}\left(1-z\right)^{j},\quad 0<z<1$$

reduces, after simplification, to the following:

$$E\left(X^{s}\right)=\beta^{s}\left[\left\{B\left(s+\delta,\delta\right)-B\left(s+\delta+1,\delta\right)\right\}+\sum_{j=0}^{\infty}\frac{\left\{\left(s+2\delta+1\right)_{j}-\left(s+2\delta\right)_{j}\right\}}{\left(\delta+1\right)_{j}}\left(\frac{1}{2}\right)^{j}\right], \tag{3.1.5}$$

where $\left(u\right)_{n}=\prod_{j=0}^{n-1}\left(u+j\right)$ denotes the Pochhammer polynomials, and $B\left(a,b\right)=\dfrac{\Gamma\left(a\right)\Gamma\left(b\right)}{\Gamma\left(a+b\right)}$ denotes the complete beta function, see, for example, Abramowitz and Stegun (6.6.2, P. 263, 1972), Gradshteyn and Ryzhik (8.392, P. 950, 2000), and Oldham et al. (18.3.1, P. 159, 58.1.2, P. 603, and 58.6.2, P. 607, 2009).

Further, noting the following results on the complete beta function and Pochhammer polynomials

$$B\left(a+1,b\right)=\frac{a}{a+b}B\left(a,b\right),$$

and

$$\left(u+1\right)_{n}=\left[1+\frac{n}{u}\right]\left(u\right)_{n},$$

see, for example, Oldham et al. (18.5.6, P. 161, and 43.13.10, P. 445, 2009), the expression (3.1.5) for the moment reduces, after simplification, to the following:

$$E\left(X^s\right)=\frac{\beta^s}{\left(s+2\delta\right)}\left[\delta B\left(s+\delta,\delta\right)+\sum_{j=0}^{\infty}\frac{j\left(s+2\delta\right)_j}{\left(\delta+1\right)_j}\left(\frac{1}{2}\right)^j\right]. \quad (3.1.6)$$

Further, by using the definition of the regularized incomplete beta function, $I_x\left(a,b\right)$, defined by

$$I_x\left(a,b\right)=\frac{B_x\left(a,b\right)}{B\left(a,b\right)},$$

see, for example, Abramowitz and Stegun (6.6.2, P. 263, 1972), Gradshteyn and Ryzhik (8.392, P. 950, 2000), and Oldham et al. (58.1.2, P. 603, 2009), we obtain, from Equation (3.1.4), the following expression for the moment, $E\left(X^s\right)$, in terms of the regularized incomplete beta function, $I_x\left(a,b\right)$, and the beta function, $B\left(a,b\right)$:

$$E\left(X^s\right)=2^{s+2\delta}\,\delta\,\beta^s\left[B\left(s+\delta,\delta\right)I_{1/2}\left(s+\delta,\delta\right)-2B\left(s+\delta+1,\delta\right)I_{1/2}\left(s+\delta+1,\delta\right)\right] \quad (3.1.7)$$

In terms of the Gauss hypergeometric function, the expression (1.7) for the moment, $E\left(X^s\right)$, is reduced to the following:

$$E\left(X^s\right)=2^{\delta}\,\delta\,\beta^s\left[\left(s+\delta\right)^{-1}{}_2F_1\left(s+\delta,1-\delta;\,s+\delta+1;\,\frac{1}{2}\right)\right.$$
$$\left.-\left(s+\delta+1\right)^{-1}{}_2F_1\left(s+\delta+1,1-\delta;\,s+\delta+2;\,\frac{1}{2}\right)\right],$$
$$=2^{\delta}\,\delta\,\beta^s\left[\left(s+\delta\right)^{-1}{}_2F_1\left(1-\delta,s+\delta;\,s+\delta+1;\,\frac{1}{2}\right)\right.$$

$$-\left(s+\delta+1\right)^{-1}{}_2F_1\left(1-\delta,\, s+\delta+1,\, s+\delta+2;\, \frac{1}{2}\right)\Bigg], \tag{3.1.8}$$

since ${}_2F_1\left(a, b; c; z\right) = {}_2F_1\left(b, a; c; z\right)$, see Abramowitz and Stegun (15.1.1, P. 556, 1972).

3.2. Integer Order Moments in Terms of Special Functions

In what follows, we derive the moments of integer order of the random variable X having the general form of J-shaped distribution with the *pdf* given by the Equation (3-0.3).

For this, considering the Equations (3..1.7) and (3.1.8), and then taking $s = n$, where $n > 0$ is an integer, we have the following expression for the nth order integer moments, $E\left(X^n\right)$, in terms of the regularized incomplete beta function, $I_x\left(a, b\right)$, and the beta function, $B\left(a, b\right)$, and the Gauss hypergeometric function, respectively.

$$\begin{aligned} E\left(X^n\right) = 2^{n+2\delta}\,\delta\,\beta^n\Big[&B\left(n+\delta, \delta\right) I_{1/2}\left(n+\delta, \delta\right) \\ &- 2B\left(n+\delta+1, \delta\right) I_{1/2}\left(n+\delta+1, \delta\right)\Big], \end{aligned} \tag{3.2.1}$$

and

$$\begin{aligned} E\left(X^n\right) = \frac{1}{2}\,\delta\,\beta^n\Big[&\left(n+\delta\right)^{-1}{}_2F_1\left(1-\delta, 1; n+\delta+1; -1\right) \\ &- \left(n+\delta+1\right)^{-1}{}_2F_1\left(1-\delta, 1; n+\delta+2; -1\right)\Big]. \end{aligned} \tag{3.2.2}$$

Using the special functions, namely, the regularized incomplete beta function, $I_x\left(a, b\right)$, the beta function, $B\left(a, b\right)$, and the Gauss hypergeometric function in the software, Maple or Mathematica, or other software, from Equations (2.1) and (2.2), one can easily compute the moments

$E\left(X^n\right)$ of different order of the random variable X by taking $n=1,2,3,\cdots$, where $n>0$ is an integer.

3.3. Explicit Expressions for the Integer Order Moments in Terms of Gamma Functions

For this, considering the general form of J-shaped distribution of the random variable X with the *pdf* given by the Equation (3.0.4), and, for convenience, replacing β and δ by b and γ, respectively, we have

$$f(x)=\frac{2\gamma}{b}\left(1-\frac{x}{b}\right)\left(\frac{x}{b}\left(2-\frac{x}{b}\right)\right)^{\gamma-1}, \quad 0\le x\le b<\infty, 0<\gamma\le 1, \tag{3.3.1}$$

and, using Equation ((3.0.4), the corresponding *cdf* is given by

$$F(x)=\left[\frac{x}{b}\left(2-\frac{x}{b}\right)\right]^{\gamma}, \quad 0\le x\le b<\infty, 0<\gamma\le 1\ . \tag{3.3.2}$$

Now, using the Equation (3.3.1), the moment of the random variable X of integer order r, where $r>0$ is an integer, is given by

$$E(X^r)=\int_0^b x^r\left(\tfrac{2\gamma}{b}(1-\tfrac{x}{b})(\tfrac{x}{b}(2-\tfrac{x}{b}))^{\gamma-1}\right)dx\,. \tag{3.3.3}$$

Let $(\frac{x}{b}(2-\frac{x}{b}))^{\gamma}=u$.

Then $u^{1/\gamma}=\frac{2x}{b}-\frac{x^2}{b^2}$,

or

$$1-u^{1/\gamma}=\left(1-\tfrac{x}{b}\right)^2,$$

which reduces to

$$x = b\left(1 - \sqrt{1 - u^{1/\gamma}}\right). \tag{3.3.4}$$

Consequently, using the above expression (3.3.4) for x in Equation (3.3.3) we have

$$E(X^r) = \int_0^1 \left(b\left(1 - \sqrt{1 - u^{1/\gamma}}\right)\right)^r du$$

$$= \int_0^1 b^r \sum_{j=0}^{r} (-1)^j \binom{r}{j} (1 - u^{1/\gamma})^{j/2} du$$

$$= b^r - b^r \sum_{j=1}^{r} (-1)^j \int_0^1 (1 - u^{1/\gamma})^{j/2} du. \tag{3.3.5}$$

Now, taking $u^{1/\gamma} = t$ and applying the beta function in terms of gamma function, that is,

$$B(a, b) = \int_0^1 (1 - t)^{a-1} t^{b-1} dt = \frac{\Gamma(a)\Gamma(b)}{\Gamma(a + b)},$$

and noting $\Gamma(a + 1) = a\Gamma(a)$, the integral in Equation (3.3.5) reduces to the following:

$$\int_0^1 (1 - u^{1/\gamma})^{j/2} du$$

$$= \int_0^1 (1 - t)^{j/2} \gamma t^{\gamma-1} dt$$

$$= \gamma B\left(\frac{j+2}{2}, \gamma\right)$$

$$= \gamma \frac{\Gamma\left(\frac{j+2}{2}\right)\Gamma(\gamma)}{\Gamma\left(\frac{j+2}{2} + \gamma\right)}$$

$$= \frac{\Gamma\left(\frac{j+2}{2}\right)\Gamma(\gamma + 1)}{\Gamma\left(\frac{j+2}{2} + \gamma\right)}.$$

Hence, using the above expression in Equation (3.3.5) the moment of the random variable X of integer order r, where $r > 0$ is an integer, is given by

$$E(X^r) = b^r - b^r \sum_{j=1}^{r} \binom{r}{j} (-1)^j \frac{\Gamma\left(\frac{j+2}{2}\right)\Gamma(\gamma+1)}{\Gamma\left(\frac{j+2}{2}+\gamma\right)}. \tag{3.3.6}$$

Thus, taking $r = 1$ and $r = 2$ in Equation (3.3.6), the first and second moments are respectively given by

$$E(X) = b - b\frac{\Gamma(\frac{3}{2})\Gamma(\gamma+1)}{\Gamma(\frac{3}{2}+\gamma)}, \tag{3.3.7}$$

and

$$E(X^2) = b^2 - 2b^2 \frac{\Gamma(\frac{3}{2})\Gamma(\gamma+1)}{\Gamma(\frac{3}{2}+\gamma)} + b^2 \frac{\Gamma(2)\Gamma(\gamma+1)}{\Gamma(\gamma+2)}$$

$$= b^2 \frac{\gamma+2}{\gamma+1} - 2b^2 \frac{\Gamma(\frac{3}{2})\Gamma(\gamma+1)}{\Gamma(\frac{3}{2}+\gamma)},$$

from which, consequently, the variance is given by

$$Var(X) = E(X^2) - [E(X)]^2 = b^2 \frac{\gamma+2}{\gamma+1} - b^2 \left[\frac{\Gamma(\frac{3}{2})\Gamma(\gamma+1)}{\Gamma(\frac{3}{2}+\gamma)}\right]^2 - b^2. \tag{3.3.8}$$

By taking different values of the parameter b in Equations (3.3.7) and (3.3.8), the moment and variance are presented in Figures 3.3.1 and 3.3.2, respectively, for $0 < \gamma < 1$.

From Figures 3.3.1 and 3.3.2, it is observed that both moment and variance are increasing functions in γ, for $0 < \gamma < 1$, when $b = 0.3, 0.5, 0.75, 1$.

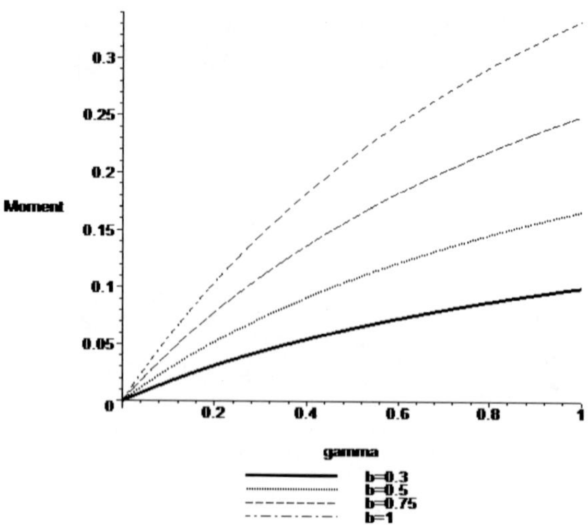

Figure 3.3.1. Plots of the Moment (3.3.7) for $0<\gamma<1$, when $b=0.3, 0.5, 0.75, 1$.

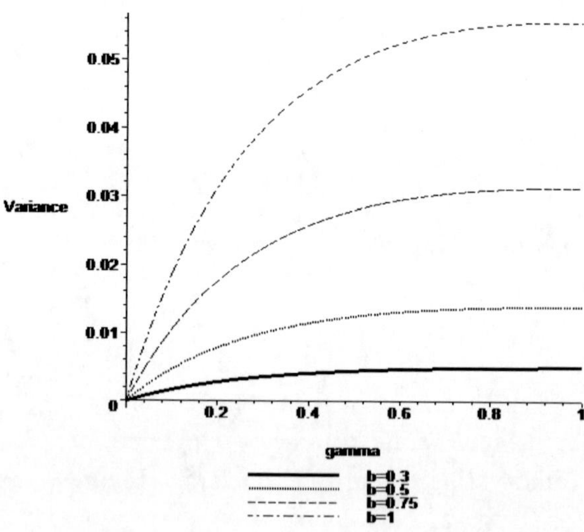

Figure 3.3.2. Plots of the Variance (3.3.8) for $0<\gamma<1$, when $b=0.3, 0.5, 0.75, 1$.

3.4. Shannon Entropy

Entropy provides an excellent tool to quantify the amount of information (or uncertainty) contained in a random observation regarding its parent distribution (population). A large value of entropy implies greater uncertainty in the data. As proposed by Shannon (1948), entropy of an absolutely continuous random variable X having the probability density function $\varphi_X(x)$ is defined as

$$H[X] = E\left[-\ln\{\varphi_X(x)\}\right] = -\int_S \varphi_X(x)\ln\{\varphi_X(x)\}\,dx\,, \quad (3.4.1)$$

where $S = \{x : \varphi_X(x) > 0\}$.

For this, considering the general form of J-shaped distribution of the random variable X with the *pdf* given by the Equation (3.0.4), and, for convenience, replacing β and δ by b and γ, respectively, we have the following expression for the *pdf* of the J-shaped distribution of the random variable X

$$f(x) = \frac{2\gamma}{b}\left(1-\frac{x}{b}\right)\left(\frac{x}{b}\left(2-\frac{x}{b}\right)\right)^{\gamma-1}, \quad 0 \le x \le b < \infty,\ 0 < \gamma < 1 \cdot \quad (3.4.2)$$

Now, using (3.0.2) in the Equation (3.4.1). Shannon entropy of the J-shaped distribution of the random variable X with the *pdf* (3.0.2), with b = 1, is given by

$$H(X) = -\int_0^1 2\gamma(1-x)\left(x(2-x)\right)^{\gamma-1}\left[\ln(2\gamma)+\ln(1-x)+(\gamma-1)\ln(x)+(\gamma-1)\ln(2-x)\right]dx$$
$$= -\int_0^1 2\gamma(1-x)\left(x(2-x)\right)^{\gamma-1}\left[\ln(2\gamma)+\ln(1-x)+(\gamma-1)\ln(x)+(\gamma-1)\ln(2-x)\right]dx,$$

and

$$H(X) = -(2\gamma)\left\{\int_0^1\left[\left(x(2-x)\right)^{\gamma-1}\left[\ln(2\gamma)+\ln(1-x)+(\gamma-1)\ln(x)+(\gamma-1)\ln(2-x)\right]-\right.\right.$$
$$\left.\left.-x^{\gamma}(2-x)^{\gamma-1}\left[\ln(2\gamma)+\ln(1-x)+(\gamma-1)\ln(x)+(\gamma-1)\ln(2-x)\right]\right]dx\,.\right.$$

Now integrating the above equation, using appropriate substitutions, and applying the relevant Equations 4.253.1/P. 538, 5.293.1/P. 557, 8.360/P. 943, and 8.370/P. 947, of Gradshteyn and Ryzhik (2000), appropriately, we obtain, after simplification, the following expression for the Shannon entropy of the of J-shaped distribution of the random variable X with the *pdf* (3.0.2), and $b = 1$:

$$H_X(\gamma) = -\ln(2\gamma) + \frac{\gamma}{2} B(\gamma, 1)[\psi(\gamma+1) - \psi(1)]$$
$$+ 2\gamma(\gamma-1)\sum_{m=0}^{\infty}\binom{\gamma-1}{m} B(\gamma, m+2)\,[\psi(\gamma+m+2) - \psi(\gamma)]$$
$$+ 2\gamma(\gamma-1)\sum_{m=0}^{\infty}(-1)^m\binom{\gamma-1}{m}\frac{1}{2m+2}\,[\beta(2m+2) - \ln(2)],$$

(3.4.3)

where $\psi(.)$ denotes the psi function, $B(a, b)$ denotes the beta function, and the function $\beta(z)$ is defined as follows:

$$\beta(z) = \frac{1}{2}\left[\psi\left(\frac{z+1}{2}\right) - \psi\left(\frac{z}{2}\right)\right],$$

see, for example, Gradshteyn and Ryzhik (8.370/P. 947, 2000), among others.

3.5. Computations of Shannon Entropy

By taking different values of the parameter γ in Equation (3.4.3), and using the software, Maple, the entropies of the J-shaped distribution of the random variable X with the *pdf* (3.4.2), when $b = 1$, are computed for some selected values of γ, which are provided in Table 3.5.1.

Furthermore, in order to discuss the shape of the entropy of the J-shaped distribution, we shall need the second derivatives of the entropy function (3.3) with respect to γ. Therefore, in view of the complicated expression for the entropy function in Equation (3.3), we have used the software, Maple, to evaluate the first and second derivatives of the entropy function (3.3) with respect to γ, and computed the respective values of its second derivative by taking different values of the parameter γ, which are also provided in Table 3.5.1.

Table 3.5.1. Computations of the Entropy, $H_X(\gamma)$, and its Second Derivative, for some selected values of the parameter γ, when $b = 1$

γ	$H_X(\gamma)$	$\frac{d^2 H_X(\gamma)}{d\gamma^2}$
0.01	-95.08117919	-0.3990064854 10^7
0.05	-16.66422786	-31641.82405
0.1	-7.326002405	-3922.742652
0.2	-2.960778703	-478.5417105
0.3	-1.645674768	-133.7851581
0.4	-1.049369996	-51.19788061
0.5	-0.7242227372	-23.02881366
0.6	-0.5266256448	-11.44788600
0.7	-0.3973053623	-6.100183018
0.8	-0.3077194814	-3.419567920
0.9	-0.2426032790	-1.988011569
0.95	-0.2163124909	-1.530567114
0.99	-0.1975602059	-1.245706989

3.6. Shapes of Shannon Entropy

The shapes of the entropy in Equation (3.0.3) are illustrated in the Figure 3.6.1, for different values of the parameter γ in the interval, $0 < \gamma < 1$, when $b = 1$.

Remark 3.6.1: From Table 3.5.1 and Figure 3.6.1, we observe the following:

a. the entropy of the J-shaped distribution is a negative, increasing, concave down function of γ, and

b. as $\gamma \to 1$, we have the entropy $H_X(\gamma) \to 0$.

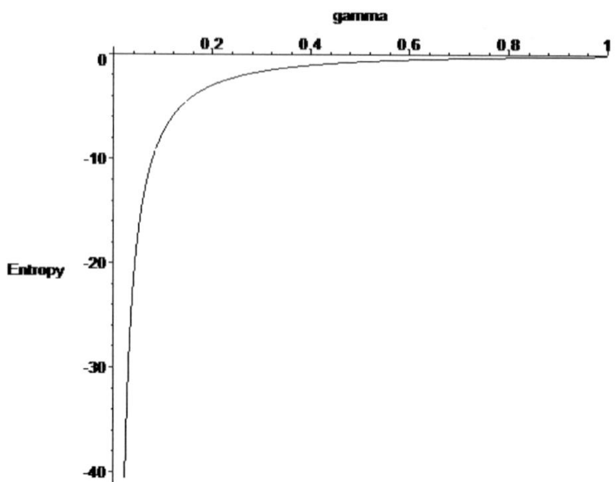

Figure 3.6.1. Plots of the Entropy (4.20) for $0 < \gamma < 1$, when $b = 1$.

3.7. Conclusion

In this chapter, we have discussed some distributional properties of the J-shaped distribution. Using the general form of the J-shaped distribution, we derived its moments. By taking different values of the scale parameter of the J-shaped distribution, the plots of the moment and variance are presented. Entropy provides an excellent tool to quantify the amount of information (or uncertainty) contained in a random observation regarding its parent distribution (population). A large value of entropy implies greater uncertainty in the data. As such, considering the general form of J-shaped distribution, its Shannon entropy is derived. Moreover, by taking different values of the parameter and using the software, Maple, the entropies of the J-shaped distribution are computed. Also, the shapes of the entropy of the J-shaped distribution are illustrated by plotting its graph for different values of the parameter. The effects of the parameters on the shapes of the entropy of the J-shaped distribution can easily be seen from its graph. We believe that the findings of this chapter would be very useful for the practitioners in various fields of studies and further enhancement of research in distribution theory, and its applications.

Chapter 4

Order Statistics

4. Introduction

In this chapter, we discuss order statistics of the J-shaped distribution and some distributional properties such as moment, variance, product moments, and covariance. Some numerical computations of these for selected values of the parameters are also as presented.

As a motivation, we first provide some preliminaries on order statistics of a random variable X in sub-section 1.1. Then, in sub-section 1.2, order statistics of the J-shaped distribution are presented.

4.1. Order Statistics

Suppose that (X_r), $(r = 1, 2, \cdots, n)$, is a sequence of n independent and identically distributed ($i.i.d.$) random variables($rv's$), each with *cdf* $F(x)$. If these are rearranged in ascending order of magnitude and written as $X_{(1;n)} \le X_{(2;n)} \le \ldots \le X_{(n;n)}$, then $X_{(r;n)}$, $(r = 1, 2, \cdots, n)$, is called the rth order statistic from a sample of size n. Further

$X_{(1;n)}$ and $X_{(n;n)}$ are called extreme order statistics, and $R = X_{(n;n)} - X_{(1;n)}$ is called the range. If $F_{(r;n)}(x)$ and $f_{(r;n)}(x)$, $(r = 1, 2, \cdots, n)$, denote the cumulative distribution function (cdf) and probability density function (pdf) of the rth order statistic $X_{(r;n)}$ respectively, then these are defined as follows:

$$F_{(r;n)}(x) = \sum_{j=r}^{n} \binom{n}{j} [F(x)]^j [1 - F(x)]^{n-j} = I_{F(x)}(r,\ n - r + 1),$$

(4.1.1)

and

$$f_{(r;n)}(x) = \frac{1}{B(r,\, n-r+1)} \left[F(x)\right]^{r-1} \left[1 - F(x)\right]^{n-r} f(x), \tag{4.1.2}$$

where $I_p(a,\, b) = \frac{1}{B(a,\, b)} \int_0^p t^{a-1} (1-t)^{b-1}\, dt$ denotes the incomplete beta function. If we take $r = 1$ and $r = n$ in (1.2), we obtain the pdf's of the extreme order statistics, that is, $X_{(1;n)}$ (minimum) and $X_{(n;n)}$ (maximum), respectively. For details on order statistics, see, for example, Ahsanullah et al. (2013), David and Nagaraja (2003), Ahsanullah and Nevzorov (2001, 2005), Arnold and Balakrishnan (1989), and Arnold et al. (2005), among others.

The joint probability density function of order statistics $X_{(1;n)}$, $X_{(2;n)}$, ... $X_{(n;n)}$ has the following form

$$f_{1,2,\ldots,n:n}(x_1, x_2, \ldots, x_n) = n! \prod_{k=1}^{n} f(x_k),\ -\infty < x_1 < x_2 < \ldots < x_n < \infty, \tag{4.1.3}$$

and

$$f_{1,2,\ldots,n:n}(x_1, x_2, \ldots, x_n) = 0, \text{ otherwise.}$$

The joint probability density function of order statistics $X_{(i;n)}$ and $X_{(j;n)}$ for $1 \le i < j \le n$ is given by

$$f_{i,j:n}(x, y) = \frac{n!}{(i-1)!(j-i-1)!(n-j)!} (F(x))^{i-1} (F(y) - F(x))^{j-i-1} (1 - F(y))^{n-j} f(x) f(y), \tag{4.1.4}$$

where $-\infty < x < y < \infty$.

We will now consider, without loss of generality, the general form of J-shaped distribution of the random variable X with the *pdf* and *cdf* given respectively by the following equations

$$f(x) = 2\gamma(1-x)\,(x(2-x))^{\gamma-1}, \quad 0 \le x \le 1, 0 < \gamma \le 1, \tag{4.1.5}$$

and

$$F(x) = \begin{cases} 0, & x < 0, \\ [x(2-x)]^{\gamma}, & 0 \le x \le 1, 0 < \gamma \le 1, \\ 1, & x > 1. \end{cases} \tag{4.1.6}$$

Now, by substituting (1.5) and (1.6) in Equation (1.2), and noting the definitions of beta and gamma functions, the *pdf* of the *rth* order statistic $X_{(r;n)}$, where $1 \le r \le n$, for a sample of size n, from the J-shaped distribution, is given by

$$\begin{aligned} f_{(r;n)}(x) &= \tfrac{1}{B(r,\, n-r+1)} \left[F(x)\right]^{r-1} \left[1 - F(x)\right]^{n-r} f(x), \\ &= \tfrac{n!}{(r-1)!(n-r)!} \left[x(2-x)\right]^{\gamma(r-1)} \left[1 - \left[x(2-x)\right]^{\gamma}\right]^{n-r} \left\{2\gamma(1-x)\,(x(2-x))^{\gamma-1}\right\}, \end{aligned} \tag{4.1.7}$$

where $0 \le x \le 1, 0 < \gamma \le 1$, and $r = 1, 2, \ldots, n$.

Taking $r = 1$ and $r = n$ in (4.1.7), we obtain the *pdf's* of the extreme order statistics, that is, $X_{(1;n)}$ (minimum) and $X_{(n;n)}$ (maximum), which are respectively given by

$$f_{(1;n)}(x) = 2\gamma n(1-x)\,(x(2-x))^{\gamma-1} \left[1 - \left[x(2-x)\right]^{\gamma}\right]^{n-1}, \tag{4.1.8}$$

and

$$f_{(n;n)}(x) = 2\gamma n(1-x)\,(x(2-x))^{\gamma n-1}, \tag{4.1.9}$$

where $0 \le x \le 1$ and $0 < \gamma \le 1$.

Using the Equation (1.4), the joint probability density function of order statistics $X_{(i;n)}$ and $X_{(j;n)}$ for $1 \le i < j \le n$, for a sample of size, n from the J-shaped distribution, is given by

$$f_{(i,j;n)}(x,y) = \frac{(4\gamma^2)(n!)}{(i-1)!(j-i-1)!(n-j)!}(1-x)(1-y)[y(2-y)]^{\gamma-1}[x(2-x)]^{\gamma i-1}.$$
$$.[1-[y(2-y)]^{\gamma}]^{n-i}[[y(2-y)]^{\gamma}-[x(2-x)]^{\gamma}]^{j-i-1}, \quad (4.1.10)$$

where $0 \le x \le 1$ and $0 < \gamma \le 1$.

4.2. Distributional Properties

In what follows, we derive the moments, product moments, variance and covariance of the rth order statistic $X_{(r;n)}$, where $1 \le r \le n$, for a sample of size n, from the J-shaped distribution, the pdf of the rth order statistic, $X_{(r;n)}$, as defined in Equation (4.1.7).

Using Equation (4.1.7), the mth moment of rth order statistic $X_{(r;n)}$, $1 \le r \le n$, is given by

$$E(X^m_{(r,n)}) = \int_0^1 \frac{n!}{(r-1)!(n-r)!} x^m [x(2-x)]^{\gamma(r-1)} [1-(x(2-x))^{\gamma}]^{n-r} [2\gamma(1-x)(x(2-x))^{\gamma-1}]dx$$

Using the transformation $u = F(x) = (x(2-x))^{\gamma}$, and noting that $F^{-1}(x) = (1-\sqrt{1-u^{1/\gamma}})$, in the above expression for $E(X^m_{(r,n)})$, we obtain

$$E(X^m_{(r,n)}) = \int_0^1 \frac{n!}{(r-1)!(n-r)!} u^{r-1}(1-u)^{n-r}[1-\sqrt{1-u^{1/\gamma}}]^m du. \quad (4.2.1)$$

In order to evaluate the integral on the right-hand side of Equation (4.2.1), we first note that

$$(1-\sqrt{1-u^{1/\gamma}})^m = u^{m/\gamma}(1+\sqrt{1-u^{1/\gamma}})^{-m},$$

and, since we have

$$(1+\sqrt{1+u})^m = \left(2^m\right)\left\{1+\tfrac{m}{1!}\tfrac{u}{(4)}+\tfrac{m(m-3)}{2!}\left(\tfrac{u}{4}\right)^2+\tfrac{m(m-4)(m-5)}{3!}\left(\tfrac{u}{4}\right)^3+\tfrac{m(m-5)(m-6)(m-7)}{4!}\left(\tfrac{u}{4}\right)^4+\cdots\right\}$$

,

where $u^2 < 1$ and m is a real number (cf. Gradshteyn and Ryzhik (2000), Equation 1.114.1, Page 21), thus

$$\begin{aligned} & u^{m/\gamma}(1+\sqrt{1-u^{1/\gamma}})^{-m} \\ & = u^{m/\gamma}[2^{-m}\{1+\tfrac{mu^{1/\gamma}}{4}+\tfrac{m(m+3)}{2!}(\tfrac{u^{1/\gamma}}{4})^2+\tfrac{m(m+3)(m+5)}{3!}(\tfrac{u^{1/\gamma}}{4})^3+.....\}] \\ & = \textstyle\sum_{j=0}^{\infty} c_j u^{(m+2j)/\gamma}, \end{aligned} \tag{4.2.2}$$

where

$$c_0 = 2^{-m}, c_1 = 2^{-(m+2)}, c_j = 2^{-(m+2j)} \tfrac{m(m+3)(m+5)\cdots(m+2i-1)}{j!}.$$

Consequently, using (2.2) in Equation (2.1), we have

$$\begin{aligned} E(X_{(r,n)}^m) &= \textstyle\int_0^1 \tfrac{n!}{(r-1)!(n-r)!} u^{r-1}(1-u)^{n-r}[1-\sqrt{1-u^{1/\gamma}}]^m du \\ &= \textstyle\int_0^1 \tfrac{n!}{(r-1)!(n-r)!} u^{r-1}(1-u)^{n-r} \sum_{j=0}^{\infty} c_j u^{(m+2j)/\gamma}. \end{aligned} \tag{4.2.3}$$

Using the definition of beta function in Equation (4.2.3), the mth moment of rth order statistic, $X_{(r;n)}$, $1 \le r \le n$, is given by

$$E(X_{(r,n)}^m) = \tfrac{n!}{(r-1)!(n-r)!} \textstyle\sum_{j=0}^{\infty} c_j B(r+(m+2j)/\delta, n-r+1), \tag{4.2.4}$$

where $c_j = 2^{-(m+2j)} \frac{m(m+3)(m+5)\cdots(m+2i-1)}{j!}$.

4.3. First and Second Moments

For this, we consider the J-shaped distribution of the random variable X with the pdf given by

$$f(x) = \frac{2\gamma}{b}\left(1-\frac{x}{b}\right)\left(\frac{x}{b}\left(2-\frac{x}{b}\right)\right)^{\gamma-1}, \quad 0 \le x \le b < \infty, 0 < \gamma \le 1,$$

and the corresponding cdf given by

$$F(x) = \begin{cases} 0, & x < 0, \\ \left[\frac{x}{b}\left(2-\frac{x}{b}\right)\right]^{\gamma}, & 0 \le x \le b < \infty, 0 < \gamma \le 1, \\ 1, & x > b. \end{cases}$$

Let $X_{(r;n)}$, where $1 \le r \le n$, denote the rth order statistic, for a sample of size n, from the above general form of J-shaped distribution. Then, the pdf of the rth order statistic $X_{(r;n)}$, where $1 \le r \le n$, for a sample of size n , from the general form J-shaped distribution, is given by

$$f_{(r;n)}(x) = \frac{1}{B(r,\, n-r+1)}\left[F(x)\right]^{r-1}\left[1-F(x)\right]^{n-r} f(x),$$

$$= \frac{n!}{(r-1)!(n-r)!}\left[\frac{x}{b}\left(2-\frac{x}{b}\right)\right]^{\gamma(r-1)}\left[1-\left[\frac{x}{b}\left(2-\frac{x}{b}\right)\right]^{\gamma}\right]^{n-r}\left\{\frac{2\gamma}{b}\left(1-\frac{x}{b}\right)\left(\frac{x}{b}\left(2-\frac{x}{b}\right)\right)^{\gamma-1}\right\}$$

where $0 \le x \le b < \infty, 0 < \gamma \le 1$, and $r = 1, 2, \ldots, n$.

Consequently, the $1st$ moment of rth order statistic $X_{(r;n)}$, $1 \le r \le n$, is given by

$$E(X_{(r,n)}) = \int_0^1 \frac{n!}{(r-1)!(n-r)!} x[F(x)]^{(r-1)}[1-F(x)]^{n-r}[f(x)]dx.$$

Then, using the transformation $u = F(x) = (x(2-x))^{\gamma}$, and noting that $F^{-1}(x) = \left(1-\sqrt{1-u^{1/\gamma}}\right)$, in the above expression for $E(X_{(r,n)})$, we obtain

$$E(X_{(r,n)}) = \frac{n!}{(r-1)!(n-r)!}\int_0^1 (b(1-\sqrt{1-u^{1/\gamma}}))u^{r-1}(1-u)^{n-r}\,du$$
$$= b - b\frac{n!}{(r-1)!(n-r)!}\int_0^1 (\sqrt{1-u^{1/\gamma}}))u^{r-1}(1-u)^{n-r}\,du.$$

Let $u^{1/\gamma} = t$, then the first moment is given by

$$E(X_{(r,n)}) = b - b\frac{n!\gamma}{(r-1)!(n-r)!}\int_0^1 (\sqrt{1-t}))t^{\gamma r-1}(1-t^{\gamma})^{n-r}\,dt$$
$$= b - b\frac{n!\gamma}{(r-1)!(n-r)!}\int_0^1 \sum_{j=0}^{n-r}\binom{n-r}{j}(-1)^j(\sqrt{1-t})t^{\gamma(r+j)-1}dt$$
$$= b - b\frac{n!\gamma}{(r-1)!(n-r)!}\sum_{j=0}^{n-r}(-1)^j\binom{n-r}{j}B(\tfrac{3}{2},\gamma(r+j)). \qquad (4.3.1)$$

Further, the second moment is given by

$$E(X^2_{(r,n)}) = \frac{n!}{(r-1)!(n-r)!}\int_0^1 (b(1-\sqrt{1-u^{1/\gamma}}))^2 u^{r-1}(1-u)^{n-r}\,du$$
$$= b^2 - 2b^2\frac{n!\gamma}{(r-1)!(n-r)!}\sum_{j=0}^{n-r}(-1)^j\binom{n-r}{j}B(\tfrac{3}{2},\gamma(r+j))$$
$$+ b^2\int_0^1 \sum_{j=0}^{n-r}(-1)^j\binom{n-r}{j}(1-t)t^{\gamma(r+j)-1}dt$$
$$= b^2 - 2b^2\frac{n!\gamma}{(r-1)!(n-r)!}\sum_{j=0}^{n-r}(-1)^j\binom{n-r}{j}B(\tfrac{3}{2},\gamma(r+j))$$
$$+ b^2\frac{n!\gamma}{(r-1)!(n-r)!}\sum_{j=0}^{n-r}(-1)^j\binom{n-r}{j}B(2,\gamma(r+j)). \qquad (4.3.2)$$

Using Equations (4.2.3) and (4.2.4), the variance of the rth order statistic, $X_{(r;n)}$, where $1 \le r \le n$, for a sample of size n, from the general form of J-shaped distribution, after simplification, is obtained as follows:

$$Var(X_{(r,n)}) = E(X^2_{(r,n)}) - (E(X_{(r,n)}))^2$$
$$= b^2\frac{n!\gamma}{(r-1)!(n-r)!}\{\sum_{j=0}^{n-r}(-1)^j\binom{n-r}{j}B(2,\gamma(r+j)) - \frac{n!\gamma}{(r-1)!(n-r)!}[\sum_{j=0}^{n-r}(-1)^j\binom{n-r}{j}B(\tfrac{3}{2},\gamma(r+j))]^2\}.$$
$$(4.3.3)$$

4.4. Product Moments

Let $X_{(r;n)}$, where $1 \le r \le n$, and $X_{(s;n)}$, where $1 \le s \le n$, denote the rth and sth order statistics, for a sample of size n, from the J-shaped distribution. For $1 \le r < s$, the product moment of the rth and sth order statistics is given by

$$E(X_{(r,n)} X_{(s,n)})$$
$$= \frac{n!}{(r-1)!(s-r-1)!(n-s)!} \int_0^1 (b(1-\sqrt{1-v^{1/\gamma}})(1-v^{n-s}) \int_0^v (b(1-\sqrt{1-u^{1/\gamma}})u^{r-1}\left(v-u\right)^{s-r-1} du\right)dv.$$
(4.4.1)

Let

$$I = \int_0^v (1-\sqrt{1-u^{1/\gamma}})u^{r-1}(v-u)^{s-r-1} du,$$

which can be expressed as

$$I = \int_0^v u^{r-1}(v-u)^{s-r-1} du - \int_0^v \sqrt{1-u^{1/\gamma}}\, u^{r-1}(v-u)^{s-r-1} du$$
$$= v^{s-1} B(r, s-r) - \sum_{j=0}^{s-r-1}$$
$$\binom{s-r-1}{j} v^{s-r-1-j} (-1)^j \int_0^v \sqrt{1-u^{1/\gamma}}\, u^{r+j-1} du$$
$$= v^{s-1} B(r, s-r) - \sum_{j=0}^{s-r-1}$$
$$\binom{s-r-1}{j} v^{s-r-1-j} (-1)^j \gamma \int_0^{v^{1/\gamma}} \left(1-t\right)^{\frac{3}{2}-1} t^{\gamma(r+j)-1} dt, \text{ where } u^{1/\gamma} = t;$$
$$= v^{s-1} B(r, s-r)$$
$$-\sum_{j=0}^{s-r-1}$$
$$\binom{s-r-1}{j} v^{s-r-1-j} (-1)^j \frac{1}{r+j} v^{r+j}\ {}_2F_1(\gamma(r+j), \frac{-1}{2}, \gamma(r+j)+1, v^{1/\gamma})$$
(4.4.2)

by using the definition of the incomplete beta function, $B_x\left(a, b\right)$, which is defined as follows

$$B_x(a,b) = \int_0^x z^{a-1}(1-z)^{b-1}\,dz = \frac{x^a}{a}\,{}_2F_1(a, 1-b; a+1; x),$$

where ${}_2F_1(a, 1-b; a+1; x)$ denotes the Gauss hypergeometric function, see, for example, Abramowitz and Stegun (6.6.1/P. 263, and 15.1/P. 556, 1972), Gradshteyn and Ryzhik (8.391/P. 950, and 9.1/P. 1039, 2000), among others.

Thus, using (4.4.2) in Equation (4.4.1), we have

$$E(X_{(r,n)}X_{(s,n)}) = \frac{n!b^2}{(r-1)!(s-r-1)!(n-s)!}\int_0^1((1-\sqrt{1-v^{1/\gamma}})(1-v)^{n-s})(v^{s-1}B(r,s-r)$$

$$-\sum_{j=0}^{s-r-1}$$

$$\binom{s-r-1}{j} v^{s-r-1-j}(-1)^j \frac{1}{r+j} v^{r+j}\,{}_2F_1(\gamma(r+j), \frac{-1}{2}, \gamma(r+j)+1, v^{1/\gamma}))\,dv$$

$$= \frac{n!b^2}{(r-1)!(s-r-1)!(n-s)!}\int_0^1 (1-v)^{n-s} v^{s-1} B(r,s-r)\,dv$$

$$- \frac{n!b^2}{(r-1)!(s-r-1)!(n-s)!}\int_0^1 \sqrt{1-v^{1/\gamma}})(1-v)^{n-s} v^{s-1} B(r,s-r)\,dv$$

$$- \frac{n!b^2}{(r-1)!(s-r-1)!(n-s)!}\left[\int_0^1 \sum_{j=0}^{s-r-1} (1-v)^{n-s}\binom{s-r-1}{j} v^{s-r-1-j}\right.$$

$$\left.\times(-1)^j \frac{1}{r+j} v^{r+j}\,{}_2F_1(\gamma(r+j), \frac{-1}{2}, \gamma(r+j)+1, v^{1/\gamma})\,dv\right]$$

$$+ \frac{n!b^2}{(r-1)!(s-r-1)!(n-s)!}\left[\int_0^1 \sum_{j=0}^{s-r-1} \sqrt{1-v^{1/\gamma}}\,(1-v)^{n-s}\binom{s-r-1}{j} v^{s-r-1-j}\right.$$

$$\left.\times(-1)^j \frac{1}{r+j} v^{r+j}\,{}_2F_1(\gamma(r+j), \frac{-1}{2}, \gamma(r+j)+1, v^{1/\gamma})\,dv\right] \tag{4.4.3}$$

Now, in order to simplify Equation (3.4.3), for the sake of simplicity, we will first evaluate different terms in Equation (4.4.3) individually as given below, and then use them back in Equation (4.4.3).

First term of Equation ((4.4.3.)

$$= \frac{n!b^2}{(r-1)!(s-r-1)!(n-s)!}\int_0^1 (1-v)^{n-s} v^{s-1} B(r,s-r)\,dv$$

$$= \frac{n!b^2}{(r-1)!(s-r-1)!(n-s)!} B(s, n-s+1)\,B(r, s-r)$$

$$= \frac{n!b^2}{(r-1)!(s-r-1)!(n-s)!} \frac{(s-1)!(n-s)!}{n!} . \frac{(r-1)!(s-r-1)!}{(s-1)!} = b^2 .$$

Second term of Equation (4.4.3)

$$= -\frac{n!b^2}{(r-1)!(s-r-1)!(n-s)!} \int_0^1 \sqrt{1-v^{1/\gamma}})(1-v)^{n-s} v^{s-1} B(r, s-r) dv .$$

But

$$\int_0^1 \sqrt{1-v^{1/\gamma}})(1-v)^{n-s} v^{s-1} dv = \gamma \sum_{j=0}^{n-s} (-1)^j \binom{n-s}{j} B(\tfrac{3}{2}, \gamma(s+j)),$$

taking $v^{1/\gamma} = t$, and using the definition of beta function, that is, $B(a, b) = \int_0^1 z^{a-1} (1-z)^{b-1} dz$.

Therefore, the Second term is reduced to the following:

Second term of Equation (4.4.3)

$$= -\frac{n!b^2\gamma}{(r-1)!(s-r-1)!(n-s)!} \frac{(r-1)!(s-r-1)!}{(s-1)!} \sum_{j=0}^{n-s} (-1)^j \binom{n-s}{j} B(\tfrac{3}{2}, \gamma(s+j))$$
$$= -\frac{n!b^2\gamma}{(n-s)!(s-1)!} \sum_{j=0}^{n-s} (-1)^j \binom{n-s}{j} B(\tfrac{3}{2}, \gamma(s+j))$$
$$= -b^2 \gamma \sum_{j=0}^{n-s} (-1)^j \frac{n!}{j!(n-s-j)!} \frac{1}{(s-1)!} B(\tfrac{3}{2}, \gamma(s+j)) .$$

Third term of Equation (4.4.3)

$$= -\frac{n!b^2}{(r-1)!(s-r-1)!(n-s)!} \int_0^1 \sum_{j=0}^{s-r-1} (1-v)^{n-s} \binom{s-r-1}{j} v^{s-r-1-j} (-1)^j \frac{1}{r+j} v^{r+j} \times ,$$
$$\times {}_2F_1(\gamma(r+j), \frac{-1}{2}, \gamma(r+j)+1, v^{1/\gamma}) dv$$

which, on taking $v^{1/\gamma} = t$, and using the Equation (7.512.5, P. 849) of Gradshteyn (1980), is reduced to the following:

Third term of Equation (4.4.3)

$$= -\tfrac{n!b^2\gamma}{(r-1)!(s-r-1)!(n-s)!}\sum_{k=0}^{n-s}\sum_{j=0}^{s-r-1}\binom{s-r-1}{j}(-1)^j\frac{1}{r+j}(-1)^k\binom{n-s}{k}B(\gamma(k+s),1)\times$$

$$\times\, {}_3F_2(\gamma(r+j),\frac{-1}{2},\gamma(k+s);\gamma(r+j)+1,\gamma(k+s)+1;1).$$

Fourth term of Equation (4.4.3)

$$= +\tfrac{n!b^2}{(r-1)!(s-r-1)!(n-s)!}\int_0^1\sum_{j-0}^{s-r-1}\sqrt{1-v^{1/\gamma}})(1-v)^{n-s}\binom{s-r-1}{j}v^{s-r-1-j}(-1)^j\frac{1}{r+j}v^{r+j},$$

$$\times\, {}_2F_1(\gamma(r+j),\frac{-1}{2},\gamma(r+j)+1,v^{1/\gamma})\,dv$$

which, on taking $v^{1/\gamma}=t$, and using the Equation (7.512.5, P. 849) of Gradshteyn (1980), is reduced to the following:

Fourth term of Equation (4.4.3)

$$= +\tfrac{n!b^2\gamma}{(r-1)!(s-r-1)!(n-s)!}\sum_{k=0}^{n-s}\sum_{j=0}^{s-r-1}\binom{s-r-1}{j}(-1)^j\frac{1}{r+j}(-1)^k\binom{n-s}{k}B(\gamma(k+s),\frac{3}{2})\times$$

$$\times\, {}_3F_2(\gamma(r+j),\frac{-1}{2},\gamma(k+s);\gamma(r+j)+\frac{3}{2},\gamma(k+s)+\frac{3}{2};1).$$

Thus using the above expressions for the first, second, third and fourth terms in Equation (2.8), the product moment, $E(X_{r,n}X_{s,n})$, of the rth and sth order statistics is reduced to the following simplified form:

$$E(X_{r,n}X_{s,n}) = b^2 - b^2\gamma\sum_{j=0}^{n-s}(-1)^j\tfrac{n!}{j!(n-s-j)!}\tfrac{1}{(s-1)!}B(\tfrac{3}{2},\gamma(s+j))$$

$$-\tfrac{n!b^2\gamma}{(r-1)!(s-r-1)!(n-s)!}\sum_{k=0}^{n-s}\sum_{j=0}^{s-r-1}\binom{s-r-1}{j}(-1)^j\frac{1}{r+j}(-1)^k\binom{n-s}{k}B(\gamma(k+s),1)\times$$

$$\times\, {}_3F_2(\gamma(r+j),\frac{-1}{2},\gamma(k+s);\gamma(r+j)+1,\gamma(k+s)+1;1)$$

$$+\tfrac{n!b^2\gamma}{(r-1)!(s-r-1)!(n-s)!}\sum_{k=0}^{n-s}\sum_{j=0}^{s-r-1}\binom{s-r-1}{j}(-1)^j\frac{1}{r+j}(-1)^k\binom{n-s}{k}B(\gamma(k+s),\frac{3}{2})\times$$

$$\times\, {}_3F_2(\gamma(r+j),\frac{-1}{2},\gamma(k+s);\gamma(r+j)+\frac{3}{2},\gamma(k+s)+\frac{3}{2};1).$$

$$= b^2 - b^2\gamma \sum_{j=0}^{n-s} (-1)^j \frac{n!}{j!(n-s-j)!} \frac{1}{(s-1)!} B(\tfrac{3}{2}, \gamma(s+j))$$

$$-\left\{ \begin{array}{l} (n! b^2 \gamma) \sum_{k=0}^{n-s} \sum_{j=0}^{s-r-1} (-1)^j \dfrac{1}{r+j} (-1)^k \dfrac{1}{j!k!(r-1)!(s-r-1-j)!(n-s-k)!} B(\gamma(k+s), 1) \times \\ \times\ {}_3F_2(\gamma(r+j), \dfrac{-1}{2}, \gamma(k+s); \gamma(r+j)+1, \gamma(k+s)+1; 1) \end{array} \right\}$$

$$+\left\{ \begin{array}{l} (n! b^2 \gamma) \sum_{k=0}^{n-s} \sum_{j=0}^{s-r-1} (-1)^j \dfrac{1}{r+j} (-1)^k \dfrac{1}{j!k!(r-1)!(s-r-1-j)!(n-s-k)!} B(\gamma(k+s), \dfrac{3}{2}) \times \\ \times\ {}_3F_2(\gamma(r+j), \dfrac{-1}{2}, \gamma(k+s); \gamma(r+j)+\dfrac{3}{2}, \gamma(k+s)+\dfrac{3}{2}; 1) \end{array} \right\}$$

(4.4.4)

Further, for a sample of size n, from the J-shaped distribution, the product of the moments of the rth order statistic, $X_{(r;n)}$, where $1 \le r \le n$, and the sth order statistic, $X_{(s;n)}$, where $1 \le s \le n$, with $1 \le r < s$, is given by

$$E(X_{r,n})E(X_{s,n}) = (b - b\tfrac{n!\gamma}{(r-1)!(n-r)!} \sum_{j=0}^{n-r} (-1)^j \tbinom{n-r}{j} B(\tfrac{3}{2}, \gamma(r+j))$$
$$.(b - b\tfrac{n!\gamma}{(s-1)!(n-s)!} \sum_{j=0}^{n-s} (-1)^j \tbinom{n-s}{j} B(\tfrac{3}{2}, \gamma(s+j))$$
$$= b^2 - b^2 \tfrac{n!\gamma}{(r-1)!(n-r)!} \sum_{j=0}^{n-r} (-1)^j \tbinom{n-r}{j} B(\tfrac{3}{2}, \gamma(r+j))$$
$$- b^2 \tfrac{n!\gamma}{(s-1)!(n-s)!} \sum_{j=0}^{n-s} (-1)^j \tbinom{n-s}{j} B(\tfrac{3}{2}, \gamma(s+j))$$
$$+ \{b^2 \tfrac{n!\gamma}{(r-1)!(n-r)!} \sum_{j=0}^{n-r} (-1)^j \tbinom{n-r}{j} B(\tfrac{3}{2}, \gamma(r+j))$$
$$\cdot \tfrac{n!\gamma}{(s-1)!(n-s)!} \sum_{j=0}^{n-s} (-1)^j \tbinom{n-s}{j} B(\tfrac{3}{2}, \gamma(s+j))\}. \qquad (4.4.5)$$

Covariance: Using (4.4.4) and (4.4.5) as given above, after simplification, for a sample of size n, from the J-shaped distribution, the covariance of the rth order statistic, $X_{(r;n)}$, where $1 \le r \le n$, and the sth order statistic, $X_{(s;n)}$, where $1 \le s \le n$, with $1 \le r < s$, is given by

$$COV(X_{r,n}, X_{s,n}) = E(X_{r,n}X_{s,n}) - E(X_{r,n})E(X_{s,n})$$

$$= \left\{ \begin{array}{l} (n!b^2\gamma)\sum_{k=0}^{n-s}\sum_{j=0}^{s-r-1}(-1)^j \dfrac{1}{r+j}(-1)^k \frac{1}{j!k!(r-1)!(s-r-1-j)!(n-s-k)!} B(\gamma(k+s), \dfrac{3}{2})\times \\ \times\ {}_3F_2(\gamma(r+j), \dfrac{-1}{2}, \gamma(k+s); \gamma(r+j)+\dfrac{3}{2}, \gamma(k+s)+\dfrac{3}{2};1) \end{array} \right\}$$

$$- \left\{ \begin{array}{l} (n!b^2\gamma)\sum_{k=0}^{n-s}\sum_{j=0}^{s-r-1}(-1)^j \dfrac{1}{r+j}(-1)^k \frac{1}{j!k!(r-1)!(s-r-1-j)!(n-s-k)!} B(\gamma(k+s),1)\times \\ \times\ {}_3F_2(\gamma(r+j), \dfrac{-1}{2}, \gamma(k+s); \gamma(r+j)+1, \gamma(k+s)+1;1) \end{array} \right\}$$

$$+ b^2 \frac{n!\gamma}{(r-1)!(n-r)!}\sum_{j=0}^{n-r}(-1)^j \binom{n-r}{j} B(\tfrac{3}{2}, \gamma(r+j))$$

$$- \left\{ b^2 \frac{n!\gamma}{(r-1)!(n-r)!}\sum_{j=0}^{n-r}(-1)^j \binom{n-r}{j} B(\tfrac{3}{2}, \gamma(r+j)) \times \frac{n!\gamma}{(s-1)!(n-s)!}\sum_{j=0}^{n-s}(-1)^j \binom{n-s}{j} B(\tfrac{3}{2}, \gamma(s+j)) \right\} \quad (4.4.6)$$

4.5. Computations of Mean, Variance and Covariance

In the following Tables 4.5.1, 4.5.2. and 4.5.3.(a, b, c), we provide means, variances and covariances of some order statistics for some selected values of the parameter γ, taking $b = 1$, without loss of generality, for a sample of size n, from the J-shaped distribution.

Table 4.5.1. Mean of the rth order statistic, $X_{(r;n)}$, of the J-shaped distribution of the random variable X, with the pdf (2.2.2), when $\boldsymbol{\beta} = 1$

$\boldsymbol{\beta}$	γ	n	r	$E(X_{(r,n)})$	$\boldsymbol{\beta}$	γ	n	r	$E(X_{(r,n)})$	$\boldsymbol{\beta}$	γ	n	r	$E(X_{(r,n)})$
1	0.3	1	1	0.1460419210	1	0.5	1	1	0.2146018365	1	0.7	1	1	0.2691429570
		2	2	0.2433176584			2	2	0.3333333333			2	2	0.3975752547
		3	3	0.3138272330			3	3	0.4109513773			3	3	0.4760350377
		4	4	0.3678669118			4	4	0.4666666666			4	4	0.5302154411
		5	5	0.4109513773			5	5	0.5091261478			5	5	0.5704853795
		6	6	0.4463253413			6	6	0.5428571428			6	6	0.6019229623
		7	7	0.4760350377			7	7	0.5704853795			7	7	0.6273430986
		8	8	0.5014415901			8	8	0.5936507936			8	8	0.6484473041
		9	9	0.5234895132			9	9	0.6134368414			9	9	0.6663322155
		10	10	0.5428571428			10	10	0.6305916306			10	10	0.6817404817

Table 4.5.2. Variance of the rth order statistic, $X_{(r;n)}$, of the J-shaped distribution of the random variable X with the pdf (2.2.2), when $\boldsymbol{\beta} = 1$

$\boldsymbol{\beta}$	γ	n	r	$Var(X_{(r,n)})$	$\boldsymbol{\beta}$	γ	n	r	$Var(X_{(r,n)})$	$\boldsymbol{\beta}$	γ	n	r	$Var(X_{(r,n)})$
1	0.3	1	1	0.0399863685	1	0.5	1	1	0.0498163913	1	0.7	1	1	0.0540832768
		2	2	0.0524318341			2	2	0.0555555555			2	2	0.0537510929
		3	3	0.0554827232			3	3	0.0530217202			3	3	0.0480413634
		4	4	0.0549532135			4	4	0.0488888889			4	4	0.0424603629
		5	5	0.0530217202			5	5	0.0447571470			5	5	0.0377394130
		6	6	0.0505872294			6	6	0.0410204082			6	6	0.0338423644
		7	7	0.0480413634			7	7	0.0377394130			7	7	0.0306183592
		8	8	0.0455571589			8	8	0.0348803224			8	8	0.0279258535
		9	9	0.0432080261			9	9	0.0323871062			9	9	0.0256521110
		10	10	0.0410204081			10	10	0.0302041232			10	10	0.0237108790

Table 4.5.3a. Covariances of the rth order statistic, $X_{(r;n)}$, and the sth order statistic, $X_{(s;n)}$, of the J-shaped distribution of the random variable X with the pdf (2.2.2), when $\boldsymbol{\beta} = 1$ and $\gamma = 0.3$

$\boldsymbol{\beta}$	γ	n	$r=1, s=2$	$r=1, s=3$	$r=1, s=4$	$r=1, s=5$	$r=1, s=6$
1	0.3	2	0.0174835347				
		3	0.0080378306	0.0088495069			
		4	0.0040683685	0.0054250639	0.0049286547		
		5	0.0022929933	0.0032925837	0.0035636633	0.0029538162	
		6	0.0014060402	0.0020987897	0.0024176803	0.0023847518	0.0018756280
		7	0.0009196868	0.0014050937	0.0016684499	0.0017506445	0.0016402684
		8	0.0006324472	0.0009809394	0.0011850867	0.0012810955	0.0012781847
		9	0.0004524688	0.0007098850	0.0008651972	0.0009520901	0.0009798068
		10	0.0003343129	0.0005279171	0.0006451756	0.0007265836	0.0007540817

Table 4.5.3b. Covariances of the rth order statistic, $X_{(r;n)}$, and the sth order statistic, $X_{(s;n)}$, of the J-shaped distribution of the random variable X with the pdf (2.2.2), when $\boldsymbol{\beta} = 1$, and $\gamma = 0.5$

$\boldsymbol{\beta}$	γ	n	$r=1, s=2$	$r=1, s=3$	$r=1, s=4$	$r=1, s=5$	$r=1, s=6$
1	0.5	2	0.0274089684				
		3	0.0169165122	0.0088495069			
		4	0.0108469996	0.0054250639	0.0113717381		
		5	0.0074360166	0.0032925837	0.0101595640	0.0029538162	
		6	0.0053863771	0.0020987897	0.0082536470	0.0023847518	0.0058851501
		7	0.0040728829	0.0014050937	0.0066623594	0.0017506445	0.0060828797
		8	0.0031849782	0.0009809394	0.0054347876	0.0012810955	0.0054941159
		9	0.0025582923	0.0007098850	0.0044962202	0.0009520901	0.0048054179
		10	0.0021002461	0.0005279171	0.0037701977	0.0007265836	0.0041759161

Table 4.5.3c. Covariances of the rth order statistic, $X_{(r;n)}$, and the sth order statistic, $X_{(s;n)}$, of the J-shaped distribution of the random variable X with the pdf (2.2.2), when $\boldsymbol{\beta} = 1$ and $\gamma = 0.7$

$\boldsymbol{\beta}$	γ	n	$r=1, s=2$	$r=1, s=3$	$r=1, s=4$	$r=1, s=5$	$r=1, s=6$
1	0.7	2	0.0333411994				
		3	0.0239999113	0.0229401939			
		4	0.0173795621	0.0203297614	0.0167385360		
		5	0.0131616767	0.0167363252	0.0165884564	0.0127500734	
		6	0.0103673735	0.0138225069	0.0147141262	0.0135701170	0.0100391514
		7	0.0084259076	0.0115879887	0.0128153981	0.0126627379	0.0112318249
		8	0.0070198175	0.0098694100	0.0111780929	0.0114539923	0.0108778196
		9	0.0059655132	0.0085269796	0.0098179008	0.0102851162	0.0101260046
		10	0.0051522523	0.0074586319	0.0086892350	0.0092401161	0.0093124870

4.6. Conclusion

In this chapter, we have discussed the order statistics of the J-shaped distribution. Some distributional properties of order statistics of the J-shaped distribution such as moment, variance, product moments, and covariance, have also been presented. Some numerical computations of these for selected values of the parameters are also provided. We believe that the findings of this chapter would be very useful for the practitioners in various fields of studies and further enhancement of research in distribution theory, and its applications.

Chapter 5

Record Values

5. Introduction

In this chapter we discuss the record values when the parent distribution is the J-shaped distribution. Some distributional properties are also presented.

An observation is called a record if its value is greater than (or analogously, less than) all the preceding observations. For example, consider the weighing of objects on a scale missing its spring. An object is placed on this scale, and its weight is measured. The 'needle' indicates the correct value but does not return to zero when the object is removed. By placing the various objects on the scale, only the weights greater than the previous ones can be recorded. These recorded weights are the record value sequence. The development of the general theory of statistical analysis of record values began with the work of Chandler (1952). Further development on record value distributions such as estimation of parameters, prediction of record values, characterizations, reconstruction of past records (which may also be viewed as the missing records) based on observed records, etc., continued with the contributions of many authors and researchers, among them Foster and Stuart (1954), Renyi (1962), Shorrock (1973), Resnick (1973), Glick (1978), Dunsmore (1983), Nagaraja (1988), Houchens (1984), Ahsanullah (1988, 1994, 1995, 2004, 2006), Samaniego and Whitaker (1986), Nevzorov (1988, 2001), Kamps (1995), Arnold et al. (1998), Rao and Shanbhag (1998), Awad and Raqab (2000), Al-Hussaini and Ahmad (2003), Gulati and Padgette (2003), Klimczak and Rychlik (2005), Ahmadi et al. (2005), Ahsanullah and Aliev (2008), Balakrishnan et al. (2009), and Ahsanullah et al. (2010), are notable.

The organization of this chapter is as follows: In Section 2, as a motivation, we provide some basic ideas, concepts and definitions of record values. Then, in Section 3, the distributions of the record values are presented. The record values of the J-shaped distribution are presented in Section 4. We provide distributional properties, moments, etc., of the record values of the J-shaped distribution in Sections 5-8. In Section 9, we have provided the percentage points associated with the *pdfs* of the distributions of upper and lower record values from the J-shaped distribution. Some concluding remarks are given in Section 10.

5.1. Basic Ideas, Definitions and Notations of Record Values

Following Ahsanullah (2004), we provide some basic ideas, definitions and notations for the distribution of record values.

Suppose that $(X_n)_{n\geq 1}$ is a sequence of $i.i.d.\ rv's$ with $cdf\ F$. Let $Y_n = \max(\min)\{X_j \mid 1 \leq j \leq n\}$ for $n \geq 1$. We say X_j is an upper (lower) record value of $\{X_n \mid n \geq 1\}$, if $Y_j > (<) Y_{j-1}, j > 1$.

By definition X_1 is an upper as well as a lower record value.

The indices at which the upper record values occur are given by the record times $\{U(n), n \geq 1\}$, where $U(n) = min$

$$U(n) = min \quad \{j \mid j > U(n-1), X_j > X_{U(n-1)}, n > 1\}$$

and $U(1) = 1$. The record times of the sequence $(X_n)_{n\geq 1}$ are the same as those for the sequence $(F(X_n))_{n\geq 1}$.

Since $F(X)$ has a uniform distribution for $rv\ X$, it follows that the distribution of $U(n), n \geq 1$ does not depend on F. We will denote $L(n)$ as the indices where the lower record values occur. By our assumption $U(1) = L(1) = 1$. The distribution of $L(n)$ also does not depend on F. For details on the lower record values, the interested readers are referred to Arnold et al. (1998), Ahsanullah (2004), and references therein.

Many properties of the record value sequence can be expressed in terms of the function

$$R(x) = -\ ln\ \overline{F}(x), 0 < \overline{F}(x) < 1,$$

which is called the cumulative hazard rate. The function given by

$$r(x) = \frac{d}{dx} R(x) = f(x)\left(\overline{F}(x)\right)^{-1},$$

is called the hazard rate. We say that F belongs to the class H^* if the hazard rate $r(x)$ is either monotone increasing or decreasing.

5.2. Distributions of Record Values

Let $F_n(x)$ and $f_n(x)$ denote the *cdf* and the *pdf*, respectively, of the nth upper record $X_{U(n)}$, for $n \geq 1$. Then, using Equations 1.2.6 and 1.2.7, Page 4, of Ahsanullah (2004), the *cdf* $F_n(x)$ and the *pdf* f_n of $X_{U(n)}$ are, respectively, given by

$$F_n(x) = \int_{-\infty}^{x} \frac{(R(u))^{n-1}}{(n-1)!} dF(u)$$

$$= \int_{-\infty}^{R(x)} \frac{u^{n-1}}{\Gamma(n)} e^{-u} du, \quad -\infty < x < \infty, \tag{5.2.1}$$

$$\bar{F}_n(x) = 1 - F_n(x) = \bar{F}(x) \sum_{j=0}^{n-1} \frac{(R(x))^j}{\Gamma(j+1)}, \quad -\infty < x < \infty, \tag{5.2.2}$$

$$= e^{-R(x)} \sum_{j=0}^{n-1} \frac{(R(x))^j}{\Gamma(j+1)}, \quad -\infty < x < \infty, \tag{5.2.3}$$

and

$$f_n(x) = \frac{(R(x))^{n-1}}{\Gamma(n)} f(x) \qquad \infty < x < \infty. \tag{5.2.4}$$

where $R(x) = -\ ln\ \overline{F}(x), 0 < \overline{F}(x) < 1$, denotes the cumulative hazard rate function of $X_{U(n)}$. Note that $F_n(x) - F_{n-1}(x) = \frac{F(x)}{f(x)} f_n(x)$.

Two $rv's$ X and Y with $cdf's$ F and G are said to be mutually symmetric if $F(x) = 1 - G(x)$ for all x, or equivalently if their corresponding $pdf's$ f and g exist, then $f(-x) = g(x)$ for all x. If a sequence of $i.i.d.$ $rv's$ are symmetric about zero, then they are also mutually symmetric about zero but not conversely. It is easy to show that for a symmetric or mutually symmetric (about zero) sequence $(X_n)_{n\geq 1}$ of $i.i.d.$ $rv's$, $X_{U(n)}$ and $X_{L(n)}$ are identically distributed. The joint pdf $f(x_1, x_2, \ldots, x_n)$ of the n record values $X_{U(1)}$, $X_{U(2)}, \ldots, X_{U(n)}$ is given by

$$f(x_1, x_2, \ldots, x_n) = \Pi_{j=1}^{n-1} r(x_j) f(x_n),$$
$$-\infty < x_1 < x_2 < \ldots < x_{n-1} < x_n < \infty, \tag{5.2.5}$$

where

$$r(x) = \frac{d}{dx} R(x) = f(x)\left(\bar{F}(X)\right)^{-1} = \frac{f(x)}{1 - F(x)}, 0 < F(x) < 1.$$

The joint pdf of $X_{U(i)}$ and $X_{U(j)}$ is

$$f_{ij}(x_i, x_j) = \frac{(R(x_i))^{i-1}}{(i-1)!} r(x_i) \frac{[R(x_j) - R(x_i)]^{j-i-1}}{(j-i-1)!} f(x_j), for -\infty < x_i < x_j < \infty.$$

In particular, for $i = 1$ and $j = n$ we have

$$f_{1n}(x_1, x_n) = r(x_1) \frac{[R(x_n) - R(x_1)]^{n-2}}{(n-2)!} f(x_n),$$
$$for -\infty < x_1 < x_n < \infty.$$

The conditional pdf of $X_{U(j)} \mid X_{U(i)} = x_i$ is

$$f(x_j \mid x_i) = \frac{f_{ij}(x_i, x_j)}{f_i(x_i)} = \frac{[R(x_i) - R(x_j)]^{j-i-1}}{(j-i-1)!} \cdot \frac{f(x_j)}{1 - F(x_i)},$$
$$for -\infty < x_i < x_j < \infty. \tag{5.2.6}$$

For $j = i+1$

$$f(x_{i+1} \mid x_i) = \frac{f(x_{i+1})}{1 - F(x_i)},$$

$$for\ -\infty < x_i < x_{i+1} < \infty. (5.2.7) \qquad (5.2.7)$$

For $i > 0, 1 \le k < m$, the joint conditional pdf of $X_{U(i+k)}$ and $X_{U(i+m)}$ given $X_{U(i)}$ is $f_{(i+k)(i+m)}\left(x, y \mid X_{U(i)} = z\right)$

$$f_{(i+k)(i+m)}\left(x, y \mid X_{U(i)} = z\right) = \frac{1}{\Gamma(m-k)} \cdot \frac{1}{\Gamma(k)} [R(y) - R(x)]^{m-k-1} \left[\frac{R(x) -}{R(z)}\right]^{k-1} \frac{f(y)\, r(x)}{\overline{F}(z)},$$

for $-\infty < z < x < y < \infty$.

The marginal pdf of the nth lower record value and the corresponding *cdf* can be derived by using the same procedure as that of the nth upper record value. For details, see, for example, Arnold et al. (1998), Ahsanullah (2004), and references therein.

Let $H(u) = -\ln F(u), 0 < F(u) < 1$

and $h(u) = -\frac{d}{du} H(u)$, then the *cdf* is expressed as

$$F_{(n)}(x) = P\left(X_{L(n)} \le x\right) = \int_{-\infty}^{x} \frac{\left(H(u)\right)^{n-1}}{(n-1)!} dF(u),$$

$$-\infty < x < \infty, \qquad (5.2.8)$$

and the corresponding pdf $f_{(n)}$ is given as

$$f_{(n)}(x) = \frac{\left(H(x)\right)^{n-1}}{(n-1)!} f(x) \quad -\infty < x < \infty. \qquad (5.2.9)$$

The joint pdf of $X_{L\,(1)}, X_{L\,(2)}, \ldots, X_{L\,(m)}$ can be written as

$$f_{(1)(2)\ldots(m)}(x_1, x_2, \ldots, x_m) = \prod_{j=1}^{m-1} h(x_j) f(x_m)$$
$$-\infty < x_m < x_{m-1} < \ldots < x_1 < \infty, = 0$$
otherwise, (5.2.10)

The joint pdf of $X_{L\,(i)}$ and $X_{L\,(j)}$ is

$$f_{(i)(j)}(x, y) = \frac{(H(x))^{i-1}}{(i-1)!} \frac{[H(y) - H(x)]^{j-i-1}}{(j-i-1)!} h(x) f(y),$$
$j > i$ and $-\infty < y < x < \infty$. (5.2.11)

Using the transformation $U = H(y)$ and $V = H(x)/H(y)$ in (5.2.10) it can easily be shown that V is distributed as $B_{i,j-i}(x)$, where $B_{m,n}(x) = B(m, n)\, x^{m-1}(1-x)^{n-1}$, and $B(m, n)$ is the Beta function. Proceeding as in the case of upper record values, we can obtain the conditional $pdf's$ of the lower record values. For record values from a family of the J-shaped distribution, the interested readers are referred to Zghoul (2011).

We consider, without loss of generality, the general form of the J-shaped distribution of the random variable X with the pdf and the cdf given by

$$f(x) = 2\gamma(1-x)\,(x(2-x))^{\gamma-1}, \quad 0 \le x \le 1, 0 < \gamma \le 1, \quad (5.2.12)$$

and

$$F(x) = \begin{cases} 0, & x < 0, \\ [x(2-x)]^{\gamma}, & 0 \le x \le 1, 0 < \gamma \le 1, \\ 1, & x > 1 \end{cases} \quad (5.2.13)$$

5.3. Probability Density and Cumulative Distribution Functions of Upper Record Values

In what follows, we first provide the distribution of upper record values when the parent distribution is the J-shaped distribution. For this, using the expressions for the cumulative hazard rate function of the J-shaped distribution, $R(x) = -\ln \bar{F}(x)$, where $0 < \bar{F}(x) = 1 - F(x) < 1$, and the corresponding expression for the *pdf* as given in (5.2.12) the *pdf* $f_n(x)$ of the upper record value $X_{U\ (n)}$ of the J-shaped distribution is given by

$$f_n(x) = \frac{(R(x))^{n=1}}{\Gamma(n)} \mathrm{f(x)}, -\infty < x < \infty.$$

$$= \frac{2\gamma(1-x)\left(x(2-x)\right)^{\gamma-1}\left(-\ln\left(1-\left[x(2-x)\right]^{\gamma}\right)\right)^{n-1}}{\Gamma(n)},$$

$$0 \le x \le 1, 0 < \gamma \le 1, \tag{5.3.1}$$

and the corresponding *cdf* $F_{(n)}(x)$ of the upper record value $X_{U\ (n)}$ of the J-shaped distribution is given by

$$F_n\ (\mathrm{x}) = (x(2-x))^{\gamma} \sum_{j=0}^{n-1} \frac{(-\ln(1-(s*2-x))^{\gamma})^{j}}{\Gamma(j+1)} \tag{5.3.2}$$

a. Plots of the *pdf* and the *cdf* of the Upper Records, for $n = 6$, and $\gamma = 0.25, 0.50, 0.75$:

The Figures 5.3.1 (a, b) illustrate the possible shapes of the plots of the *pdf* $f_n(x)$, and the *cdf*, $F_n(x)$, respectively, of the upper record value $X_{U\ (n)}$ of the J-shaped distribution, for some selected values of the parameters, namely, for $n = 6$, and $\gamma = 0.25, 0.50, 0.75$:

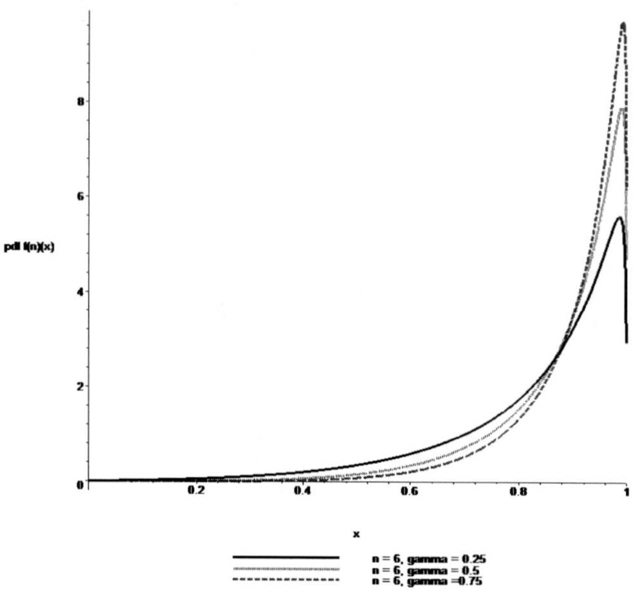

Figure 5.3.1. (a): PDFs of $X_{U(n)}$, when $n = 6$, and $\gamma = 0.25, 0.50, 0.75$. SOLID: $\gamma = 0.25$, DOT: $\gamma = 0.5$, and DASH: $\gamma = 0.75$.

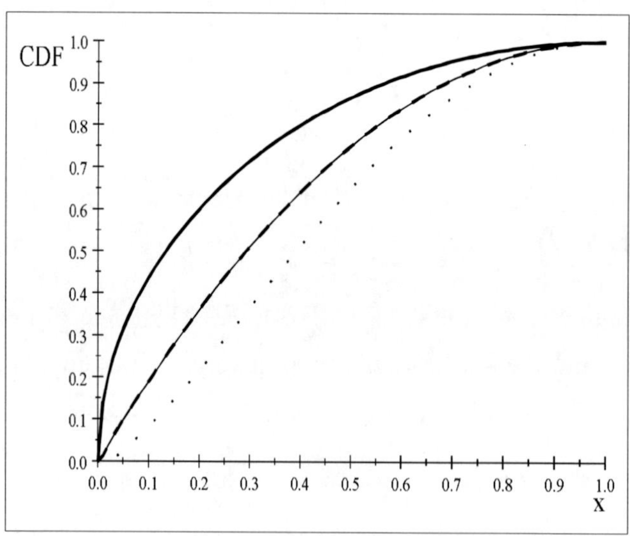

Figure 5.3.1. (b): *CDFs* of $X_{U(n)}$, when $n = 6$, and $\gamma = 0.25, 0.50, 0.75$ .SOLID: $\gamma = 0.25$, DOT: $\gamma = 0.5$, and DASH: $\gamma = 0.75$.

From the plots in Figure 5.3.1 (b), the effects of the parameters on the behaviors of the corresponding shapes of the plots of the *cdf*, $F_n(x)$,, of the upper record value $X_{U\ (n)}$ of the J-shaped distribution are easily observed. For example, "the increasing, and concave down shape behaviors of the *cdf* (5.2.13), $F_n(x)$, are evident from the curves in the above-said plot 5.3.1 (b).

b. Plots of the *pdf* and the *cdf* of the Upper Records, for $\gamma = 1$, and $n = 3, 5, 10$.

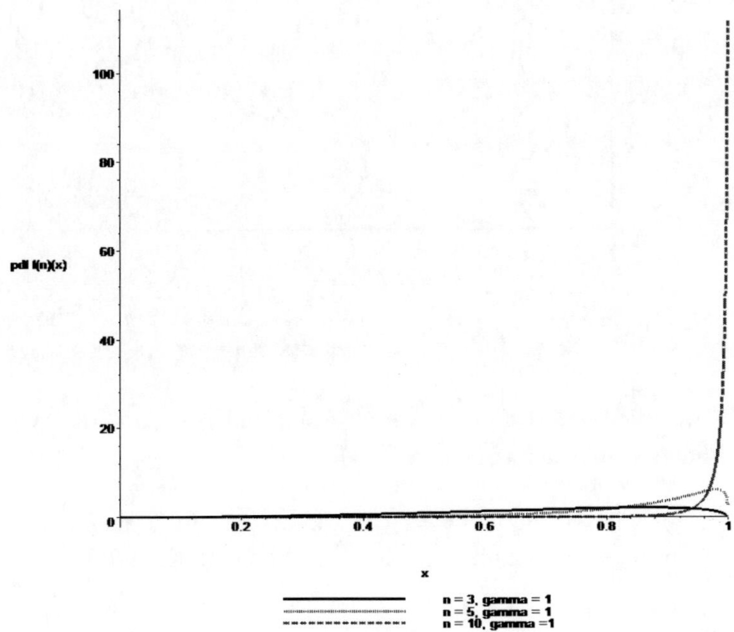

Figure 5.3.2. (a): *PDFs* of $X_{U\ (n)}$, when $\gamma = 1$, and $n = 3, 5, 10$. SOLID: $n = 3$, DOT: $n = 5$, and DASH: $n = 10$.

Figure 5.3.2 (a, b) illustrate the possible shapes of the plots of the *pdf* $f_n(x)$, and the *cdf*, $F_n(x)$, (respectively, of the upper record value $X_{U\ (n)}$ of the J-shaped distribution for some selected values of the parameters, namely, for $\gamma = 1$, and $n = 3, 5, 10$.

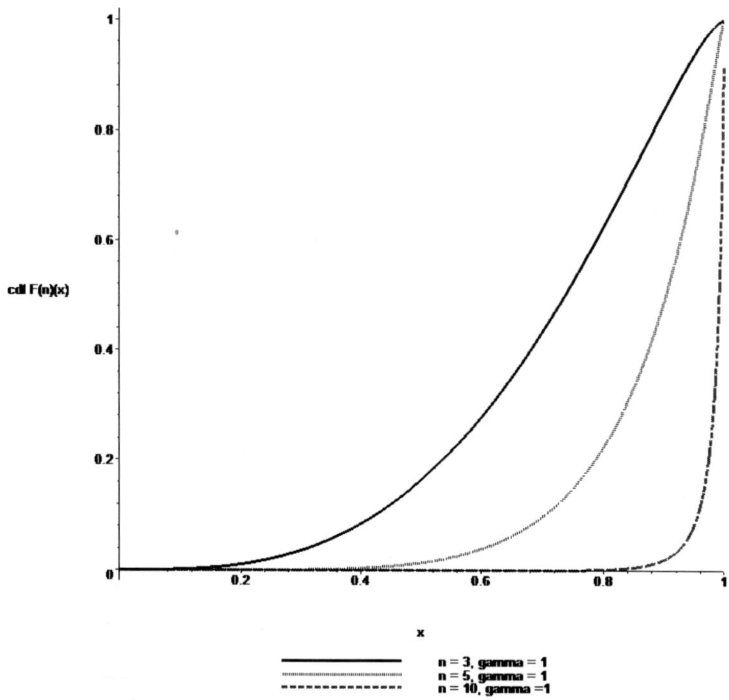

Figure 5.3.2. (b): *CDFs* of $X_{U\ (n)}$, when $\gamma = 1$, and $n = 3, 5, 10$. SOLID: $n = 3$, DOT: $n = 5$, and DASH: $n = 10$.

From the plots in Figure 5.3.2 (b), the effects of the parameters on the behaviors of the corresponding shapes of the plots of the *cdf* (5.2.13), $F_{(n)}(x)$, of the upper record value $X_{U\ (n)}$ of the J-shaped distribution are easily observed. For example, "the increasing and concave up shape behaviors of the *cdf* (5.4.2) $F_n(x)$ $estimator$re evident from the curves in the above-said plot 5.3..2 (b).

5.4. Probability Density and Cumulative Distribution Functions of Lower Record Values

Following Ahsanullah (2004, Pages 6 - 7), the pdf and the cdf of the nth lower record value, $X_{L(n)}$, $n \geq 1$, of the J-shaped distribution can be derived by using the same procedure as that of the nth upper record value as discussed before. Thus, Ahsanullah (2004, Pages 6 - 7), the *pdf* of nth lower records $X_{L(n)}$ of the J-shaped distribution can be expressed as follows:

$$f_{(n)}(x)=\frac{2\gamma(1-x)\,(x(2-x))^{\gamma-1}\left(-\ln\left(\left[x(2-x)\right]^{\gamma}\right)\right)^{n-1}}{\Gamma(n)},$$

$$0\le x\le 1, 0<\gamma\le 1. \tag{5.4.1}$$

The corresponding cdf is given by

$$F_{(n)}(x)=P\left(X_{L(n)}\le x\right)=\int_0^x\frac{\left(H(u)\right)^{n-1}}{(n-1)!}dF(u)$$

$$=\int_0^x\frac{\left(-\ln F(u)\right)^{n-1}}{(n-1)!}dF(u)$$

$$=\int_0^{F(x)}\frac{\left(-\ln t\right)^{n-1}}{(n-1)!}dt\text{ , where } F(u)=t$$

$$=\frac{1}{(n-1)!}\int_{-\ln F(x)}^{\infty}e^{-u}\,u^{n-1}\,du\text{ , taking } -\ln t=u$$

$$=\frac{1}{(n-1)!}\Gamma\left(n,-\ln F(x)\right)$$

$$=\frac{1}{\Gamma(n)}\Gamma\left(n,-\ln\left[x(2-x)\right]^{\gamma}\right),\; 0\le x\le 1, 0<\gamma\le 1, \tag{5.4.2}$$

where $\Gamma(a,z)=\int_z^{\infty}e^{-u}\,u^{a-1}\,du$ denotes the incomplete gamma function.

By direct differentiation, it can be easily seen that

$$\frac{dF_{(n)}(x)}{dx}=\frac{1}{\Gamma(n)}\frac{d\left(\Gamma\left(n,-\ln\left[x(2-x)\right]^{\gamma}\right)\right)}{dx}$$

$$=\frac{2\gamma(1-x)\left(x(2-x)\right)^{\gamma-1}\left(-\ln\left(\left[x(2-x)\right]^{\gamma}\right)\right)^{n-1}}{\Gamma(n)}$$

$$=f_{(n)}(x).$$

Remark 5.5.1: An Alternative Expression for the *CDF* of Lower Record Values

We will consider the i.i.d. sequence $\{X_n, n \geq 1\}$ with *cdf* F(x) as

$$F(x) = (x(2-x))^{\gamma},\ 0 \leq x \leq 1,\ 0 < \gamma \leq 1. \tag{5.4.3}$$

Let $f_{(n)}(x)$ be the *pdf* of the n-th lower record $X_{L(n)}$, then

$$f_{(n)}(x) = \frac{2\gamma^n}{\Gamma(n)}(1-x)(x(2-x))^{\gamma-1}(-(\ln(x(2-x))))^{n-1}, \tag{5.5.4}$$

where $0 \leq x \leq 1,\ 0 < \gamma \leq 1$.

The corresponding *cdf* $F_{(n)}(x)$ is

$$F_{(n)}(x) = F(x)\sum_{j=0}^{n-1}\frac{(-\ln F(x))^j}{\Gamma(j+1)}$$
$$= (x(2-x))^{\gamma}\sum_{j=0}^{n-1}\frac{(-\gamma(\ln(x(2-x))))^j}{\Gamma(j+1)}. \tag{5.4.5}$$

a. Plots of the *pdf* and the cdf of the Lower records, for $n=3$, and $\gamma = 0.4, 0.6, 0.85$. The following Figures 5.4.1 (a, b) illustrate the possible shapes of the plots of the *pdf* $f_{(n)}(x)$, and the *cdf* (5.4.5), $F_{(n)}(x)$, respectively, of the lower record value $X_{U(n)}\, X_{L(n)}$ of the J-shaped distribution, for some selected values of the parameters, namely, for $n=3$, and $\gamma = 0.4, 0.6, 0.85$.

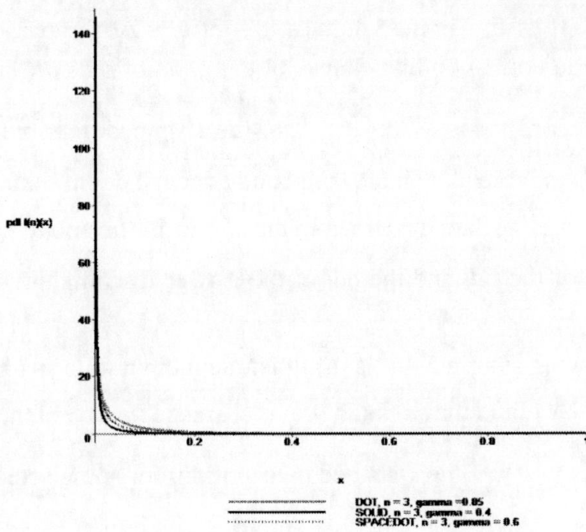

Figure 5.4.1. (a): *PDFs* of $X_{L(3)}$: SOLID: $\gamma = 0.4$, SPACEDOT: $\gamma = 0.6$, and DOT: $\gamma = 0.85$.

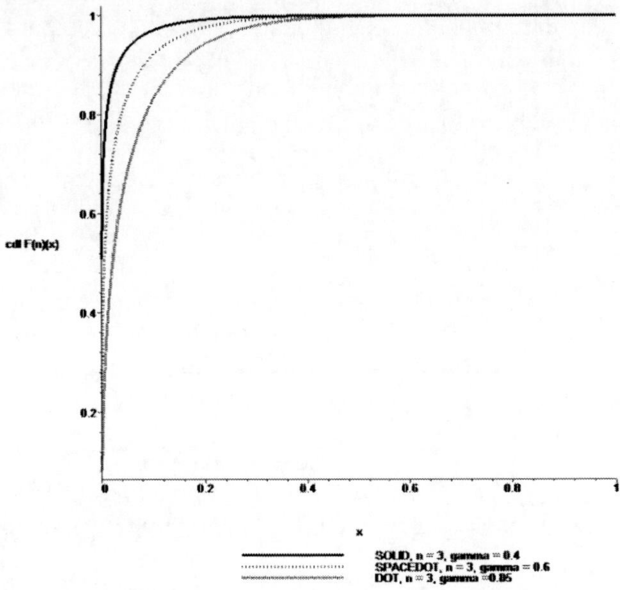

Figure 5.4.1. (b): *CDFs* of $X_{L(3)}$: SOLID: $\gamma = 0.4$, SPACEDOT: $\gamma = 0.6$, and DOT: $\gamma = 0.85$.

From the plots in Figure 5.4.1 (b), the effects of the parameters on the behaviors of the corresponding shapes of the plots of the *cdf* (5.5.4), $F_{(n)}(x)$, of the lower record value $X_{U\,(n)}\,X_{L(n)}$ of the J-shaped distribution are easily observed. For example, "the increasing, and concave down shape behaviors of the *cdf* (4.6), $F_{(n)}(x)$, are evident from the curves in the above-said plot 5.5.2.

(b). Plots of the *pdf* and the *cdf* of the Lower Records, for $\gamma = 0.5$, **and** $n = 1, 3, 5$.

The following Figure 5.4.2 (a, b) illustrate the possible shapes of the plots of the *pdf* $f_{(n)}(x)$, and the *cdf* 5.4.5 $F_{(n)}(x)$, respectively, of the lower record value $X_{U\,(n)}\,X_{L(n)}$ of the J-shaped distribution, for some selected values of the parameters, namely, for $\gamma = 0.5$, and $n = 1, 3, 5$:

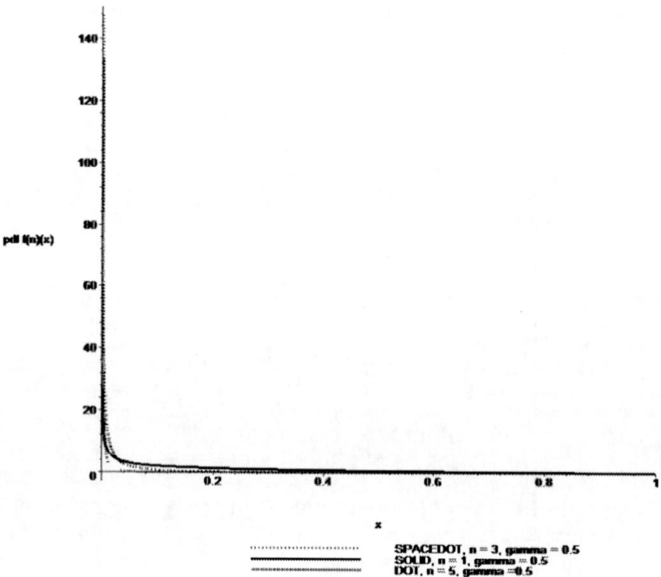

Figure 5.4.2. (a): *PDFs* for $\gamma = 0.5$, SOLID: $n = 1$, SPACEDOT: $n = 3$, and DOT: $n = 5$.

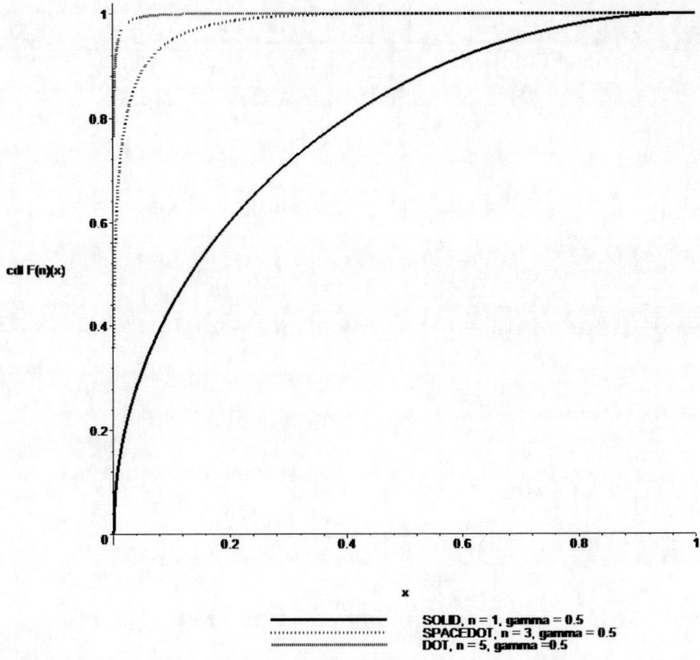

Figure 5.4.2. (b): *CDFs* for $\gamma = 0.5$, SOLID: $n = 1$, SPACEDOT: $n = 3$, and DOT: $n = 5$.

From the plots in Figure 5.4.2 (b), the effects of the parameters on the behaviors of the corresponding shapes of the plots of the (*cdf*), $F_{(n)}(x)$, of the lower record value $X_{U(n)} X_{L(n)}$ of the J-shaped distribution are easily observed. For example, "the increasing and concave down Shape behaviors of the *cdf* (5.6.1), $F_{(n)}(x)$, are evident from the curves in the above-said plot 5.4.1 (b).

5.5. Probability Density Functions of Joint and Conditional Record Values

Let $f_{(m,n)}(x, y)$ be the joint *pdf* of the mth and nth lower records $X_{L(m)}$ and $X_{L(n)}$, then

$$f_{(m,n)}(x,y) = \frac{2\gamma^n}{\Gamma(m)\Gamma(n-m)} \cdot \frac{x(1-x)}{x(2-x)} \left(-\gamma(lnx + \ln(2-x))\right)^{m-1}$$
$$\times (lny + \ln(2-y)) - (lnx + \ln(2-x)))^{n-m-1}(1-$$
$$y)(y(2-y))^{\gamma-1}.$$

Let $f_{(n|m)}(y|x)$ be the conditional *pdf* of $X_{L(n)} \mid X_{L(m)} = x$, then $f_{(n|m)}(y|X_{L(m)} = x)$

$$= \frac{2\gamma^{n-m}}{\Gamma(n-m)} [(lny + \ln(2-y)) - (lnx + \ln(2-x))]^{n-m-1} \frac{(1-y)(y(2-y))^{\gamma-1}}{(x(2-x))^{\gamma}}.$$

If n = m + 1, then

$$f_{(m+1|m)}(y|X_{L(m)} = x)$$
$$f_{(m+1|m)}(y|X_{(m)} = x\} = \frac{2\gamma(1-y)(y(2-y))^{\gamma-1}}{(x(2-x))^{\gamma}}.$$

Further, let $f_{m,n}(x,y)$ be the joint *pdf* of the mth and nth upper records, $X_{U(m)}$ and $X_{(m,n)}$), then

$$f_{(m,n)}(x,y) = \frac{4\gamma^2}{\Gamma(m)\Gamma(n-m)} (\ln(1 - (x(2-x))^{\gamma}))^{m-1}$$
$$.(\ln(1-(x(2-x))^{\gamma}) - \ln(1-(y(2-y))^{\gamma}))^{n-m-1}$$
$$.\frac{(1-x)}{x(2-x)}(1 - y)(y(2-y))^{\gamma-1}.$$

The conditional *pdf* of $X_{U(n)}|X_{U(m)} = x$ is then given by

$$f_{(m,n)}(y|X_{U(m)} = x)$$
$$= \frac{2\gamma}{\Gamma(n-m)}$$
$$(\ln(1-(x(2-x))^{\gamma}) - \ln(1-(y(2-y))^{\gamma}))^{n-m-1}$$
$$\times \frac{(1-y)(y(2-y))^{\gamma-1}}{1-(x(2-x))^{\gamma}}.$$

If n = m+1, then

$$f_{(m+1|m)}(y|X_{U(m)} = x) = \frac{2\gamma(1-y)(y(2-y))^{\gamma-1}}{1-(x(2-x))^{\gamma}}.$$

5.6. Moments of Record Values

In this section, the moments of the lower and upper record values of the J-shaped distribution are presented.

a. Moments of the Lower Record Values

Theorem 5.6.1. For r = 1, 2, ... , and n ≥ 1, let $\mu_{(n)}^{r}$ $\mu_{(n)}^{r}$ be the rth moment of the nth lower record $X_{L(n)}$, with the *pdf* given by the Equation (5.5.4), that is,

$$f_{(n)}(x) = \frac{2\gamma(1-x)\left(x(2-x)\right)^{\gamma-1}\left(-\ln\left(\left[x(2-x)\right]^{\gamma}\right)\right)^{n-1}}{\Gamma(n)},$$

$0 \leq x \leq 1, 0 < \gamma \leq 1.$

Then

$$\mu_{(n)}^{r} = \sum_{j=0}^{\infty}(-1)^{j} a_{j}\left(\frac{\gamma}{\gamma+r+j}\right)^{n},$$

where

$a_0 = (2)^{-r}$, $a_1 = r(2)^{-r-2}$, and

$$a_{j} = \frac{(2)^{-r-2j}(r)(r+2j-1)}{j!\,(r+j)!}, \quad j \geq 2.$$

Proof: We have

$\mu_{(n)}^{r} = \int_0^1 x^r \frac{2\gamma^n}{\Gamma(n)}(1 - x)$
$(x(2-x))^{\gamma-1}(-(lnx + \ln(2-x)))^{n-1}$ dx.

Let $x(2 - x) = u$. Then

$$\mu_{(n)}^{r} = \int_0^1 \frac{\gamma^n}{\Gamma(n)}(1-\sqrt{(1-u)})^r u^{\gamma-1}(-lnu)^{n-1}\text{du}.$$

Since $\left(1+\sqrt{1-u}\right)^r\left(1-\sqrt{1-u}\right)^r = u^r$, we have

$$\left(1-\sqrt{1-u}\right)^r = u^r\left(1+\sqrt{1-u}\right)^{-r}.$$

Using the above expression, we have

$$\mu_{(n)}^{r} = \int_0^1 \frac{\gamma^n u^{r+\gamma-1}}{\Gamma(n)}(1+\sqrt{(1-u)})^{-r}(-lnu)^{n-1}\text{du}.$$

Now, the following binomial series expansion for the expression $(1+\sqrt{1+u})^r$ is well-known:

$$(1+\sqrt{1+u})^r = \left(2^r\right)\left\{1+\frac{r}{1!}\frac{u}{(4)}+\frac{r\,(r-3)}{2!}\left(\frac{u}{4}\right)^2+\frac{r\,(r-4)(r-5)}{3!}\left(\frac{u}{4}\right)^3+\frac{r\,(r-5)(r-6)(r-7)}{4!}\left(\frac{u}{4}\right)^4+\cdots\right\}$$

where $u^2 < 1$ and r is a real number (cf. Gradshteyn and Ryzhik (2000), Equation 1.114.1, Page 21), which can easily be expressed as

$$(1+\sqrt{1+u})^r = \sum_{j=0}^{\infty} b_j\,(u)^j,$$

where

$$b_0 = (2)^r,\ b_1 = r(2)^{r-2},\text{ and}$$

$$b_j = \frac{(2)^{r-2j}(r)}{j!}\prod_{i=1}^{j-1}(r-j-i),\quad j \geq 2.$$

Thus, using the binomial expression for $\left(1+\sqrt{1-u}\right)^{-r}$, and simplifying, we easily obtain

$$\mu_{(n)}^{r} = \sum_{j=0}^{\infty} (-1)^{j} a_{j} \left(\frac{\gamma}{\gamma + r + j} \right)^{n},$$

where

$a_0 = (2)^{-r}$, $a_1 = r(2)^{-r-2}$, and

$$a_j = \frac{(2)^{-r-2j}(r)(r+2j-1)}{j!\,(r+j)!}, \quad j \geq 2.$$

This completes the proof of Theorem 5.8.1.

Theorem 5.6.2. For r = 1, 2,... . and n ≥ 1, let $\mu_{(n)}^{r}$ be the rth moment of the nth lower record $X_{L(n)}$. Then we have the following recurrence relation:

$$(1+\frac{2\gamma}{r+1})\mu_{(n)}^{r+1} = 2(1+\frac{\gamma}{r})\mu_{(n)}^{r} + 2\gamma(\frac{1}{r+1}\mu_{(n-1)}^{r+1} - \frac{1}{r}\mu_{(n-1)}^{r})$$

Proof: We have

$$2\mu_{(n)}^{r} - \mu_{(n)}^{r+1} =$$

$$\frac{1}{\Gamma(n)} \int_0^1 x^r (2-x)\left(-lnF(x)\right)^{n-1} f(x)dx$$

$$= \frac{2\gamma}{\Gamma(n)} \int_0^1 x^{r-1}(1-x)\left(-lnF(x)\right)^{n-1} F(x)dx$$

$$= \frac{2\gamma}{\Gamma(n-1)} \int_0^1 (\frac{x^r}{r} - \frac{x^{r+1}}{r+1})\left(-lnF(x)\right)^{n-2} f(x)dx$$

$$- \frac{2\gamma}{\Gamma(n)} \int_0^1 (\frac{x^r}{r} - \frac{x^{r+1}}{r+1})\left(-lnF(x)\right)^{n-1} f(x)dx$$

$$= 2\gamma(\frac{1}{r}\mu_{(n-1)}^{r} - \frac{1}{r+1}\mu_{(n-1)}^{r+1}) - 2\gamma(\frac{1}{r}\mu_{(n)}^{r} - \frac{1}{r+1}\mu_{(n)}^{r+1}).$$

On simplification we have our result.

b. Moments of the Upper Record Values

Theorem 5.6.3. Let μ_n^r be the rth moment of the nth upper record value, $X_{U(n)}$. Then, for r = 1, 2, … and n ≥ 1, we have

$$\mu_n^r = \sum_j^\infty a_j(-1)^j \sum_{k=0}^\infty \binom{\frac{\gamma+j}{\gamma}-1}{k} (-1)^k (k+1)^{-n},$$

where

$$a_0 = 2^{-r}, a_1 = r2^{-r-2},$$

and

$$a_j = \frac{2^{-r-2j}(r)\,(r+2j-1)!}{j!\,(r+j)!}, \text{j} \geq 2.$$

Proof: We have

$$\mu_n^r = \int_0^1 2(\gamma x^r\,(-\ln\,(1-(x(2-x))^\gamma)\,)^{\,n-1}$$
$$.(1-\text{x})\,(x(2-x))^{\gamma-1}\text{dx}.$$

Using u = x (2 - x), we obtain

$$\mu_n^r = \frac{\gamma}{\Gamma(n)}\int_0^1 u^{\gamma-1}\left(1-\sqrt{(1-u)}\right)^r(-\ln\,(1-u^\gamma)^n\,\text{du}$$
$$= \frac{\gamma}{\Gamma(n)}\int_0^1 u^{\gamma-1}u^r\left(1+\sqrt{(1-u)}\right)^{-r}(-\ln\,(1-u^\gamma)^n$$

$$= \frac{\gamma}{\Gamma(n)}\sum_{j=0}^\infty \int_0^1 a_j(-1)^j\,u^{\gamma+r+j-1`}(-\ln\,(1-u^\gamma)^n,$$

or, taking $u^\gamma = w$, we have

$$\mu_n^r =$$
$$\frac{1}{\Gamma(n)}\sum_{j=0}^\infty \int_0^1 a_j\,(-1)^j(1-w)^{\frac{\gamma+r+j-1}{\gamma}}(-\ln \text{w})^n(1-w)^{\frac{1-\gamma}{\gamma}}$$
$$\text{dw}$$
$$=$$
$$\frac{1}{\Gamma(n)}\sum_{j=0}^\infty \int_0^1 a_j\,(-1)^j\sum_{k=0}^\infty \binom{\frac{r+\gamma+j}{\gamma}-1}{k}(-1)^k w^k(-\ln \text{w})^n$$

dw

$$= \sum_{j=0}^{\infty} a_j(-1)^j \sum_{k=0}^{\infty} \binom{\frac{r+\gamma+j}{\gamma}-1}{k} (-1)^k (k+1)^{-n},$$

where

$$a_0 = 2^{-r}, a_1 = r2^{-r-2},$$

and

$$a_j = \frac{2^{-r-2j}(r)(r+2j-1)!}{j!(r+j)!}, j \geq 2.$$

This completes the proof of Theorem 5.8.3.

Remark 5.6.1. For further discussions on the moments, including the recurrence relations for the moments, the maximum likelihood estimator (MLE), unbiased and consistent estimator, and the minimum variance unbiased estimator (MVUE) of the parameters of the J-shaped distribution, etc., the interested readers are referred to Zghoul (2011) and references therein. For moments of kth lower record values from the J-shaped distribution, see Kumar (2014).

5.7. Percentage Points

As described in Chapter 1, the shapes of probability distributions in many fields of research, such as, biology, computer science, control theory, economics, engineering, genetics, hydrology, medicine, number theory, statistics, physics, psychology, reliability, risk management, etc., exhibit J-shaped distributions. Therefore, the percentage points of the distribution of the record values when the parent distribution is the J-shaped distribution would be also very useful for the practitioners in various fields of studies and further enhancement of research in distribution theory, and its applications. As such, in this section, motivated by the importance of the percentage points, in order to demonstrate the applications of our results, we have tabulated the percentage points associated with the *pdfs* of the distributions of upper and lower record values from the J-shaped distribution.

For any α, where $0 < \alpha < 1$, the $(100\alpha)th$ percentile or the quantile of order α of a continuous probability distribution with the *pdf* $f_X(x)$ is a number t_α such that the area under $f_X(x)$ to the left of t_α is α. That is, the percentage points t_α are any roots of the equation

$$F(t_\alpha) = \int_{-\infty}^{t_\alpha} f_X(u)\, du = \alpha,$$

or,

$$t_\alpha = F^{-1}(\alpha),$$

where $F(t_\alpha)$ denotes the corresponding *cdf* of the continuous probability distribution under consideration. Thus, the percentage points t_α of said continuous probability distribution can be computed by numerically solving the equations for the *cdf* $F(t_\alpha)$, that is, $t_\alpha = F^{-1}(\alpha)$, for any α, where $0 < \alpha < 1$, by taking different sets of values of the parameters involved.

Using the software, Maple, we have computed the percentage points of the upper and lower record values from the J-shaped distributions for some selected values of the parameters, which are provided in Tables 5.10.1-5.10.4. In a similar way, we can compute the percentage points for other values of the parameters.

Table 5.7.1. Percentiles of the Upper Record Values of the Random Variable X with the *pdf* (5.4.1), when $n = 6$, and $\gamma = 0.25, 0.50, 0.75$

n	γ	75%	80%	85%	90%	95%	99%
6	0.25	0.951133	0.961620	0.971400	0.980635	0.989574	0.997153
	0.50	0.965440	0.972854	0.979775	0.986306	0.992630	0.997990
	0.75	0.971780	0.977834	0.983490	0.988820	0.99400	0.998360

Table 5.7.2. Percentiles of the Upper Record Values Random Variable X with the *pdf* (5.4.1), when $n = 3, 5, 10$, and $\gamma = 1$

γ	n	75%	80%	85%	90%	95%	99%
1	3	0.859170	0.882300	0.905724	0.930133	0.957058	0.985049
	5	0.956597	0.965282	0,973580	0.981626	0.989712	0.996980
	10	0.997413	0.998088	0.998673	0.999180	0.999612	0.999917

Table 5.7.3. Percentiles of the Lower Record Values of the Random Variable X with the *pdf* (5.5. 4), when $n = 3$, and $\gamma = 0.40, 0.60, 0.85$

n	γ	75%	80%	85%	90%	95%	99%
3	0.40	0.006684	0.010832	0.018122	0.032322	0.066984	0.185248
	0.60	0.028510	0.039494	0.055996	0.083120	0.137410	0.281308
	0.85	0.067830	0.085850	0.110615	0.147640	0.213955	0.366518

Table 5.7.4. Percentiles of the Lower Record Values of the Random Variable X with the *pdf* (5.5. 4) when $n = 1, 3, 5$, and $\gamma = 0.5$

γ	n	75%	80%	85%	90%	95%	99%
0.5	1	0.338563	0.400000	0.473220	0.564111	0.687751	0.858933
	3	0.015927	0.023485	0.035562	0.056786	0.102714	0.237162
	5	0.000594	0.001037	0.001907	0.003863	0.009769	0.039502

5.8. Conclusion

In this chapter, we have discussed the distribution of record values when the parent distribution is a J-shaped distribution. The probability density and cumulative distribution functions of upper and lower record values, along with their graphs, are provided. We have also provided the probability density functions of joint and conditional record values when the parent distribution is a J-shaped distribution. Furthermore, some distributional properties, moments of the lower and upper record values and the recurrence relations for moments are presented. Finally, in order to demonstrate the applications of our results, we have tabulated the percentage points associated with the *pdfs* of the distributions of upper and lower record values from the J-shaped distribution. We hope that the findings of this chapter will be useful for the practitioners in various fields of studies and further enhancement of research in record value theory and its application.

Chapter 6

Sum, Product and Ratio

6. Introduction

Modeling and analyzing of real lifetime data play very important roles in many fields of research such as biology, computer science, control theory, economics, engineering, genetics, hydrology, medicine, number theory, statistics, physics, psychology, reliability, risk management, etc., among others. In view of this, therefore, in recent years, there has been a growing interest in modeling non-normal or skew distributions and to make the well-known distributions more flexible. It has been observed that the enhanced models developed as such would be more applicable and suitable for certain lifetime data when usual models are not. Thus, various systems of non-normal or skew distributions have been constructed to provide approximations to wide variety of distributions, see, for example, Patel et al. (1976), Johnson et al. (1994), Balakrishnan and Nevzorov (2003), and Forbes et al. (2011), among others. For further discussions on some recently developed distributions, the interested readers are referred to Shakil and Ahsanullah (2012), Tahir and Nadarajah (2015), Tahir and Cordeiro (2016), de Brito et al. (2019), and references therein. These systems are designed with the requirements of ease of computation and feasibility of algebraic manipulation. To meet the requirements, there must be as few parameters as possible in defining a member of the system as such the new model could be very useful to capture non-Gaussian behavior of the data.

Furthermore, the distributions of the sum, product, and ratio of two independent random variables arise in many fields of research, for example, automation, biology, computer science, control theory, economics, engineering, fuzzy systems, genetics, hydrology, medicine, neuroscience, number theory, statistics, physics, psychology, reliability, risk management, etc., see, for example, Grubel (1968), Rokeach and Kliejunas (1972), Springer (1979), Kordonski and Gertsbakh (1995, I & II), Ladekarl et al. (1997), Amari and Misra (1997), Sornette (1998), Cigizoglu and Bayazit (2000), Brody et al. (2002), and Galambos and Simonelli (2005), among others.

Since the distributions of the sum, product, and ratio of two independent random variables are of great importance in many problems of research in applied sciences and arise in many fields of pure and applied research, this has increased the need to explore more statistical results on the sum, product, and ratio of independent random variables. As pointed out in Chapter 1 of this book, the J-shaped distribution is one of the most important continuous probability distributions, and has been extensively studied and widely used by many researchers in the fields of probability, statistics and other applied sciences since it was introduced. In this chapter, we present the distributions of the sum, product, and ratio of two independently distributed J-shaped random variables.

Many researchers and authors have studied the distributions of the sum $|X + Y|$, product $|XY|$, and ratio $\left|\frac{X}{Y}\right|$, when X and Y are independent random variables and belong to the same and different families, which are reviewed and discussed in the following paragraphs:

6.1. Literature Review

As pointed out by Nadarajah (2008), "sums of exponential random variables, sums of gamma random variables, sums of lognormal random variables and the sums of Rayleigh random variables all have pivotal roles in wireless communications and related areas". For more details and applications to practical examples, the interested readers are referred to Nadarajah (2008), and references therein. For further examples on the sum $|X + Y|$, when X and Y are independent random variables and belong to the same family, the interested readers are referred to Nadarajah (2005a), Nagar and Ramirez-Vanegas (2013), Ahsanullah et al. (2014), and Mallick et al. (2018), among others.

The distribution of the product $|XY|$ of two independent random variables X and Y belonging to the same family is also an important area of research both from theoretical and applications point of view. It arises naturally in many problems of applied sciences. As pointed out by Shakil et al. (2008), "the distributions of the product |XY|, when X and X are independent random variables, arise in many applied problems of biology, economics, engineering, genetics, hydrology, medicine, number theory, order statistics, physics, psychology, etc.". For more details, the interested readers are referred to

Shakil et al. (2008), and references therein. For further examples on the product $|XY|$, when X and Y are independent random variables and belong to the same family, the interested readers are referred to Nadarajah (2005a, b), Nagar and Ramirez-Vanegas (2013), Ahsanullah et al. (2014), Mallick et al. (2018), and Adamska et al. (2022), among others.

The distribution of the ratio $\left|\frac{X}{Y}\right|$ of two independent random variables X and Y belonging to the same family is also one of the important research areas both from theoretical and applications point of view. It arises naturally in many applied problems of biology, economics, engineering, genetics, hydrology, medicine, number theory, order statistics, physics, psychology, etc. Some of the notable examples are the ratios of inventory in economics, ratios of inheritance in genetics, ratios of target to control precipitation in meteorology, ratios of mass to energy in physics, among others. As pointed out by Nadarajah and Gupta (2006), "the distribution of the ratio $Z = \left|\frac{X}{Y}\right|$ is of interest in biological and physical sciences, econometrics, and ranking and selection. Examples include Mendelian inheritance ratios in genetics, mass to energy ratios in nuclear physics, target to control precipitation in meteorology, and inventory ratios in economics. Another important example is the stress–strength model in the context of reliability. It describes the life of a component which has a random strength Y and is subjected to random stress X. The component fails at the instant that the stress applied to it exceeds the strength and the component will function satisfactorily whenever $Y > X$. Thus, $\Pr(X < Y)$ is a measure of component reliability. It has many applications especially in engineering concepts such as structures, deterioration of rocket motors, static fatigue of ceramic components, fatigue failure of aircraft structures and the aging of concrete pressure vessels". For further contributions on the distributions of the ratio $\left|\frac{X}{Y}\right|$, when X and Y are independent random variables and belong to the same family, we refer to Marsaglia (1965), Lee et al. (1979), Korhonen and Narula (1989), Press (1969), Pham-Gia (2000), Nadarajah (2005b), Nadarajah and Gupta (2005, 2006), Nadarajah and Kotz (2006, 2007), Ali et al (2007), Khoolenjani and Khorshidian (2009), and Ahsanullah et al. (2014). The interested readers are also referred to Nadarajah (2010), who studied the distributional properties as well as estimation of the ratio of independent Weibull random variables, to Shakil and Ahsanullah (2011), who studied the ratio of two independent Rayleigh random variables and the distributional properties of the corresponding record values of the said ratio, and to Nagar and Ramirez-Vanegas (2013) for the quotient of independent non-central beta type 3 random variables. We also refer to Gunduz and Genc (2015) obtained the exact distributions of the quotients $\frac{X}{Y}$ and $\frac{Y}{X+Y}$ when X and Y are independent and triangularly distributed random variables, and pointed out that "these quotients are useful especially in operations research and reliability engineering." For more details and some reliability applications of the results, we refer to Gunduz and Genc (2015). The interested readers are also referred to Mallick et al. (2018) for the ratio of Kumaraswami random variables, and Shakil et al. (2022) for the ratio of Lindley random variables.

6.2. Distributions of $|X+Y|$, $|XY|$, and $\left|\frac{X}{Y}\right|$, When X and Y Belong to Different Families

The distributions of the sum $X+Y$, product $|XY|$, and ratio $\left|\frac{X}{Y}\right|$, when X and Y are independent random variables and belong to different families, are also of considerable importance and current interest. These have been studied by many researchers and authors.

For the distributions of the sum $X+Y$, when X and Y are independent random variables and belong to different families, see, for example, Nadarajah (2005) for the linear combination, product and ratio of normal and logistic random variables, Nadarajah and Kotz (2005) for the linear combination of exponential and gamma random variables, Nadarajah and Kotz (2006) for the linear combination of logistic and Gumbel random variables, Nason (2006) for the sum of t and Gaussian random variables, Kibria and Nadarajah (2007) for the linear combination of exponential and Rayleigh random variables, Shakil and Kibria (2009) for linear combination of gamma and Rayleigh random variables, and Nadarajah and Kotz (2011) for the linear combination of normal and Laplace random variables.

For the distributions of the product $|XY|$ and ratio $\left|\frac{X}{Y}\right|$, when X and Y are independent random variables and belong to different families, the interested readers are referred to Nadarajah (2005c), Nadarajah and Ali (2004, 2005), Ali and Nadarajah (2004, 2005), Nadarajah and Kotz (2005, 2006, 2011), Shakil and Kibria (2006, 2007), Shakil et al. (2006, 2008), Garg and Agrawal (2007), Hassan et al. (2020), Obeid and Kadry (2019, 2021, 2022), and references therein.

6.3. Some Preliminaries on Distributions of the Sum $|X+Y|$, Product $|XY|$, and Ratio $\left|\frac{X}{Y}\right|$

For the sake of completeness, this section presents some preliminaries and basic definitions of the distributions of the sum $|X+Y|$, product $|XY|$, and ratio $\left|\frac{X}{Y}\right|$, when X and Y are independent random variables and belong to the same family. For details, the interested readers are referred to Lukacs (1972), Dudewicz and Mishra (1988), Rohatgi and Saleh (2001), Kapadia et al. (2005), and Larson and Marx (2006), among others.

Let X and Y be any two independent, absolutely continuous random variables with *pdfs* $f_X(x)$ and $f_Y(y)$, respectively. Note that the sum $|X+Y|$, product $|XY|$, and ratio $\left|\frac{X}{Y}\right|$ of X and Y are also random variables, (for details, see Lukacs (1972), among others).

Let X and Y be two continuous independent random variables with density functions $f(x)$ and $g(y)$ respectively. Assume that $f(x)$ and $g(y)$ are defined for all real numbers.

Let $Z = X + Y$ for $-\infty < X,\ Y < +\infty$. Then, the sum $Z = X + Y$ is a continuous random variable with the *cdf*, $F_Z(z)$, and *pdf*, $f_Z(z)$, which are respectively given by

(i) $F_Z(z) = P(Z \le z) = P(X + Y \le z) = \int_{-\infty}^{+\infty} f_X(x) F_Y(z - x)\,dx$. (6.3.1)

(ii) $f_Z(z) = \int_{-\infty}^{+\infty} f_X(x) f_Y(z - x)\,dx$. (6.3.2)

(iii) Convolution: The probability density function $f_Z(z)$ in (ii) above is also called the convolution of the *pdfs* $f_X(x)$ and $f_Y(y)$, which is expressed as $\{f_Z(z)\} = \{f_X(z)\} * \{f_Y(z)\}$.

a. Definition (Distribution of a linear combination). Let $Z = aX + bY$ for $-\infty < X,\ Y < +\infty$, and $a, b \in \Re - \{0\}$. Then, the *cdf* and Distribution of a Linear Combination *pdf* of the random variable Z are respectively given by

$$F_Z(z) = \Pr(aX + bY \le z) = \Pr\left(X \le \frac{z - bY}{a}\right)$$

$$= \int_{-\infty}^{z/b} F_X\left(\frac{z - by}{a}\right) f_Y(y)\,dy \quad (6.3.3)$$

$$f_Z(z) = \int_{-\infty}^{z/b} f_X\left(\frac{z - by}{a}\right) f_Y(y)\,dy \quad (6.3.4)$$

b. Distribution of a Product $-\infty < X,\ Y < +\infty$.

Let $W = XY$ Then

$$F_W(w) = P(W \le w) = P(XY \le w) = P\left(X \le \frac{w}{Y}\right)$$

$$= \begin{cases} \int_0^{\infty} F_X\left(\frac{w}{y}\right) f_Y(y)\,dy, & Y > 0 \\ \int_0^{\infty} F_X\left(\frac{w}{y}\right) f_Y(y)\,dy + F_Y(0) - \int_{-\infty}^{0} F_X\left(\frac{w}{y}\right) f_Y(y)\,dy, & Y < 0 \end{cases} \quad (6.3.5)$$

$$f_W(w) = = \begin{cases} \int_0^{\infty} \frac{1}{y} f_X\left(\frac{w}{y}\right) f_Y(y)\,dy, & Y > 0, -\infty < w < \infty \\ \int_{-\infty}^{\infty} \left|\frac{1}{y}\right| f_X\left(\frac{w}{y}\right) f_Y(y)\,dy, & Y < 0, -\infty < w < \infty \end{cases} \tag{6.3.6}$$

c. Distribution of a Ratio:

Let $U = \dfrac{X}{Y}$ for $-\infty < X,\ Y < +\infty$. Then

$$F_U(u) = P(U \le u) = P\left(\frac{X}{Y} \le u\right) = P(X \le uY) \tag{6.3.7}$$

$$= \begin{cases} \int_0^{\infty} F_X(u\,y) f_Y(y)\,dy, & Y > 0 \\ \int_0^{\infty} F_X(u\,y) f_Y(y)\,dy + F_Y(0) - \int_{-\infty}^{0} F_X(u\,y) f_Y(y)\,dy, & Y < 0 \end{cases}$$

$$f_U(u) = = \begin{cases} \int_0^{\infty} y\, f_X(u\,y) f_Y(y)\,dy, & Y > 0, -\infty < u < \infty \\ \int_{-\infty}^{\infty} |y|\, f_X(u\,y) f_Y(y)\,dy, & Y < 0, -\infty < u < \infty \end{cases} \tag{6.3.8}$$

Let X and Y be any two independent, absolutely continuous random variables with *pdfs* $f_X(x)$ and $f_Y(y)$, respectively. Suppose that the first and second moments of both X and Y exist. Then

$$\mathrm{E}(X + Y) = \mathrm{E}(X) + \mathrm{E}(Y)$$
$$\mathrm{E}(XY) = \mathrm{E}(X)\mathrm{E}(Y)$$
$$Var(X + Y) = Var(X) + Var(Y)$$

Note that the above results can be extended to any finite sums or products of independently distributed random variables. For details, see Lukacs (1972), and Dudewicz & Mishra (1988), among others.

6.4. Distribution of the Sum, Product and Ratio of Independent J-Shaped Random Variables

This section presents the distributions of the sum, product and ratio of independent J-shaped random variables. Let X and Y be any two independent, absolutely continuous random variables having the general form of the J-shaped distribution with the *pdfs* respectively given by

$$f(x)=\frac{2\alpha}{a}\left(1-\frac{x}{a}\right)\left(\frac{x}{a}\left(2-\frac{x}{a}\right)\right)^{\alpha-1}, \quad 0\le x\le a<\infty,\ 0<\alpha\le 1, \tag{6.4.1}$$

and

$$f(y)=\frac{2\beta}{b}\left(1-\frac{y}{b}\right)\left(\frac{y}{b}\left(2-\frac{y}{b}\right)\right)^{\beta-1}, \quad 0\le y\le b<\infty,\ 0<\beta\le 1. \tag{6.4.2}$$

The corresponding *cdfs* are respectively given by

$$F(x)=\begin{cases}0, & x<0,\\ \left[\frac{x}{a}\left(2-\frac{x}{a}\right)\right]^{\alpha}, & 0\le x\le a<\infty,\ 0<\alpha\le 1,\\ 1, & x>a,\end{cases} \tag{6.4.3}$$

and

$$F(y)=\begin{cases}0, & x<0,\\ \left[\frac{y}{b}\left(2-\frac{y}{b}\right)\right]^{\beta}, & 0\le y\le b<\infty,\ 0<\beta\le 1,\\ 1, & x>b.\end{cases} \tag{6.4.4}$$

Then, in what follows, adopted from Zhou et al. (2006), for the sake of completeness, we provide, without proofs, the exact *pdfs* of the distributions of the sum $R=X+Y$, product $P=XY$ and ratio $W=\frac{X}{X+Y}$, in Theorems 3.1, 3.2 and 3.3, respectively. To demonstrate the application of their results, they also tabulated the percentage points associated with the *pdfs* of the distributions of $R=X+Y$ and $W=\frac{X}{X+Y}$, which, for the sake of completeness, we have also provided. As pointed out by Zhou et al. (2006), "if X and Y represent the failure time and the repair/replacement time, respectively, then the sum $R=X+Y$ represents the inter-arrival time between failures, whereas the ratio $W=\frac{X}{X+Y}$ represents the efficiency of the repair / replacement scheme". For more details of the motivation, discussions, applicability and practical examples, the interested readers are referred to Zhou et al. (2006).

(a) Distribution of the Sum $R=X+Y$

Theorem 6.4.1, Suppose X and Y are independent random variables from the J-shaped family with *pdfs* (6.4.1) and (6.4.2) respectively. Then, Zhou et al. (2006) derives the exact *pdf* of the distribution of the sum $R=X+Y$, for $0<r<a+b$, as follows:

$$f_R(r)=\frac{4\alpha\beta}{a^{2\alpha}b^{2\beta}}\sum_{i=0}^{\infty}\sum_{j=0}^{\infty}\binom{\alpha-1}{i}\binom{\beta-1}{j}(-1)^{i+j}(2a)^{\alpha-1-i}(2b)^{\beta-1-j}I_{ij}(r), \tag{6.4.5}$$

where $I_{ij}(r)$ is given by:

$$I_{ij}(r)=\begin{cases} abB(\alpha+i,\beta+j)-arB(\alpha+i,\beta+j+1) \\ \qquad -brB(\alpha+i+1,\beta+j)+r^2B(\alpha+i+1,\beta+j+1), \quad \text{if } r\le a \\ abB_{a/r}(\alpha+i,\beta+j)-arB_{a/r}(\alpha+i,\beta+j+1) \\ \qquad -brB_{a/r}(\alpha+i+1,\beta+j)+r^2B_{a/r}(\alpha+i+1,\beta+j+1), \quad \text{if } a<r\le b \\ abB_{a/r}(\alpha+i,\beta+j)-abB_{1-b/r}(\alpha+i,\beta+j)-arB(\alpha+i,\beta+j+1) \\ \qquad +arB_{1-b/r}(\alpha+i,\beta+j+1)-brB_{a/r}(\alpha+i+1,\beta+j) \\ \qquad +brB_{1-b/r}(\alpha+i+1,\beta+j)+r^2B_{a/r}(\alpha+i+1,\beta+j+1) \\ \qquad -r^2B_{1-b/r}(\alpha+i+1,\beta+j+1), \quad \text{if } r>b \end{cases}$$

where

$$B(a,b)=\int_0^1 t^{a-1}(1-t)^{b-1}\,dt=\frac{\Gamma(a)\Gamma(b)}{\Gamma(a+b)}=B(b,a),$$

and

$$B_v(p,q)=\int_0^v t^{p-1}(1-t)^{q-1}\,dt=\frac{v^p}{p}\,{}_2F_1(p,1-q;p+1;v),$$

denote the beta and the incomplete beta functions, respectively, and $\frac{v^p}{p}\,{}_2F_1(p,1-q;p+1;v)$ denotes the Gauss hypergeometric function; cf.8.380.1, page 948, 8.391, page 950, and 9.1, page 1039, Gradshteyn and Ryzhik (2000), and 6.6.1, page 263, and 15.1 page 556, Abramowitz and Stegun (1972).

For proof, see Zhou et al. (2006).

Remark 6.4.1: According to Zhou et al. (2006), "if *X* and *Y* represent the failure time, then the sum $R=X+Y$ represents the inter-arrival time between failures."

(b) Distribution of the Product $P=XY$

Theorem 6.4.2. The *pdf* of the distribution of the product $P=XY$, for $0<p<ab$, is given by: y Zhou et al. (2006) as,

$$f_P(p)=\frac{2^{\alpha+\beta}\alpha\beta}{a^{\alpha}b^{\beta}}\sum_{i=0}^{\infty}\sum_{j=0}^{\infty}\binom{\alpha-1}{i}\binom{\beta-1}{j}\frac{(-1)^{i+j}}{(a)^{i}(b)^{j}}(p)^{\beta+j-1}I_{ij}(p), \quad (6.4.6)$$

where $I_{ij}(p)$ is given by:

$$I_{ij}(p)=\left(1+\frac{p}{ab}\right)\delta(\alpha-\beta+i-j)-\frac{1}{a}\delta(\alpha-\beta+i-j+1)-\frac{p}{b}\delta(\alpha-\beta+i-j-1)$$

where $\delta(.)$ is given by:

$$\delta(m)=\begin{cases}\dfrac{a^{m}-\left(\dfrac{p}{b}\right)^{m}}{m}, & \text{if } m\neq 0\\ \log\left(\dfrac{ab}{p}\right), & \text{if } m=0\end{cases}.$$

(c). Distribution of the Ratio $W=\dfrac{X}{X+Y}$

Theorem 6.4.3. Zhou et al. (2006), has obtained the exact *pdf* of the distribution of ratio $W=\dfrac{X}{X+Y}$, for $0<w<1$, as follows:

$$f_W(w)=\frac{4\alpha\beta}{a^{2\alpha}b^{2\beta}}\sum_{i=0}^{\infty}\sum_{j=0}^{\infty}\binom{\alpha-1}{i}\binom{\beta-1}{j}(-1)^{i+j}, \\ \times(2a)^{\alpha-1-i}(2b)^{\beta-1-j}(w)^{\alpha+i-1}(1-w)^{\beta+j-1}I_{ij}(w) \quad (6.4.7)$$

where $I_{ij}(w)$ is given by:

$$I_{ij}(w)=\begin{cases}\dfrac{ab}{(\alpha+\beta+i+j)}\left(\dfrac{b}{1-w}\right)^{(\alpha+\beta+i+j)} \\ -\dfrac{a(1-w)+bw}{(\alpha+\beta+i+j+1)}\left(\dfrac{b}{1-w}\right)^{(\alpha+\beta+i+j+1)} \\ +\dfrac{w(1-w)}{(\alpha+\beta+i+j+2)}\left(\dfrac{b}{1-w}\right)^{(\alpha+\beta+i+j+2)}, & if\ w<\dfrac{a}{a+b} \\ \dfrac{ab}{(\alpha+\beta+i+j)}\left(\dfrac{a}{w}\right)^{(\alpha+\beta+i+j)} \\ -\dfrac{a(1-w)+bw}{(\alpha+\beta+i+j+1)}\left(\dfrac{a}{w}\right)^{(\alpha+\beta+i+j+1)} \\ +\dfrac{w(1-w)}{(\alpha+\beta+i+j+2)}\left(\dfrac{a}{w}\right)^{(\alpha+\beta+i+j+2)}, & if\ w\geq\dfrac{a}{a+b}\end{cases}.$$

Proof: For proof, see Zhou et al. (2006).

Remark 6.4.4. As pointed out by Zhou et al. (2006), "if X and Y represent the repair/replacement time, then the ratio $W=\dfrac{X}{X+Y}$ represents the efficiency of the repair / replacement scheme".

6.5. Conclusion

The distributions of the sum, product, and ratio of two independently distributed continuous random variables X and Y, when they belong to the same family, are of great importance in many problems of research in applied sciences. In this chapter, following Zhou et al. (2006), we have introduced the distributions of the sum, product, and ratio of two independently distributed J-shaped random variables. The distributions of the sum $|X+Y|$, product $|XY|$, and ratio $W=\dfrac{X}{X+Y}$, when X and Y are independent random variables and have the J-shaped distributions, have been reviewed. The expressions for the *pdfs* as proposed by Zhou et al. (2006) are presented. Since the percentage are useful for the researchers in many fields of research, the intended readers are referred to Zhou et al (2006for the tabulated values of of the percentage associated with the *pdfs* of the distributions of R = X+Y and $W = \frac{X}{X+Y}$.

We believe that the findings of this chapter would be very useful for the practitioners in various fields of studies and further enhancement of research in distribution theory, and its applications.

Chapter 7

Characterizations

7. Introduction

The problems of characterizations of both discrete and continuous probability distributions have been studied by many authors and researchers. As pointed out by Nagaraja (2006), "A characterization uses a certain distributional or statistical property of a statistic or statistics that uniquely determines the associated stochastic model." On the other hand, Koudou and Ley (2014) point out that, "In probability and statistics, a characterization theorem occurs when a given distribution is the only one which satisfies a certain property. Besides their evident mathematical interest per se, characterization theorems also deepen our understanding of the distributions under investigation and sometimes open unexpected paths to innovations which might have been uncovered otherwise." Thus, it is very important to characterize a particular probability distribution subject to certain conditions first before applying it to some real-world data. According to Glänzel (1990), since a characterization of a particular probability distribution states that it is the only distribution that satisfies some specified conditions, these characterizations may serve as a basis for parameter estimation. Also, see Glänzel et al. (1984), and Glänzel (1987). Furthermore, Glänzel (1990) points out that the characterizations by truncated moments may also be useful in developing some goodness-of-fit tests of distributions by using data whether they satisfy certain properties given in the characterizations of distributions. These conditions are used by various authors to test goodness of fit, efficiency of a particular test of hypothesis and the power of a particular estimating, etc. For example, Volkova and Nikitin (2015) used a well-known characterization result of Ahsanullah (1978) to test exponentiality of a distribution. For more on the goodness-of-fit and symmetry tests based on the characterization properties of distributions, the interested readers are referred to recent nice papers of Nikitin (2017), Miloševic (2017), and Akbari [(2020), and references therein.

A probability distribution can be characterized through various methods. The development of the general theory of the characterizations of probability distributions by truncated moment began with the work by Kagan et al. (1973), followed by Galambos and Kotz (1978). Since the truncated distributions arise in practical statistics where the ability of record observations is limited to a given threshold or within a specified range, there has been a great interest, in recent years, in the characterizations of probability distributions by truncated moments. As such, further development on the characterizations of probability distributions by truncated moments continued with the contributions of many authors and researchers, among them Kotz and Shanbhag (1980), Glänzel et al. (1984), and Glänzel (1987) are notable. Since it is necessary to confirm whether the given continuous probability distribution satisfies the underlying requirements by its characterization before a particular probability distribution model is applied to fit the real-world data, therefore the characterizations of probability distributions by truncated moments may play an important part in the determination of distributions by using certain properties in the given data. For example, as pointed out by Barr and Sherrill (1999), in a variety of applications, the mean and variance of normal distributions that have been truncated in various ways are needed. Similarly, as pointed out by Kim and Jeon (2013), in actuarial science, the credibility theory proposed by Buhlmann (1967) allows actuaries to estimate the conditional mean loss for a given risk to establish an adequate premium to cover the insured's loss. In their paper, Kim and Jeon (2013) have proposed a credibility theory based on truncation of the loss data, or the trimmed mean, which also contains the classical credibility theory of Buhlmann (1967), as a special case.

Most of the characterizations, as stated above, are based on a simple relationship between two different moments truncated from the left at the same point. However, on the other hand, another important method of characterization, as appears in the literature, is the method of left and right truncated moments, by taking a product of reverse hazard rate and another function of the truncated point, see, for example, Ahsanullah et al. (2014), Shakil et al. (2018), and Ahsanullah (2017).

Motivated by the importance of the practicality of the characterizations of lifetime distribution as discussed above, in this chapter, we provide the proposed characterizations of the distribution of the J-shaped distribution, by truncated moment, by considering a product of reverse hazard rate and another function of the truncated point. We also investigate characterizations of the J-shaped distribution by order statistics and upper record values.

The organization of this chapter is as follows: In Section 7.1, we have derived the probability density function of the J-shaped distribution as a solution of the generalized Pearson differential equation. In Section 7.2, we have presented some characterizations of the J-shaped distribution by truncated first moment. Section 7.3 contains the characterizations by order statistics. In Section 7.4, we have presented the characterizations of the J-shaped distribution by record values.

7.1. As a Solution of the Generalized Pearson Differential Equation

Without loss of generality, we will consider the general form of the J-shaped distribution. An absolutely continuous random variable X is said to have the general form of the J-shaped distribution if its distribution and probability density functions are, respectively, given by

$$F(x)=\begin{cases}0, & x<0,\\ \left[\frac{x}{\beta}\left(2-\frac{x}{\beta}\right)\right]^{\gamma}, & 0\le x\le\beta<\infty, 0<\gamma\le 1,\\ 1, & x>\beta,\end{cases} \tag{7.1.1}$$

and

$$f(x)=\frac{2\gamma}{\beta}\left(1-\frac{x}{\beta}\right)\left(\frac{x}{\beta}\left(2-\frac{x}{\beta}\right)\right)^{\gamma-1}. \tag{7.1.2}$$

Theorem 7.1.

The probability density function (7.2) of the general form of the J-shaped distribution is a solution of the following generalized Pearson differential equation:

$$\frac{f'(x)}{f(x)}=\frac{a_0+a_1x+a_2x^2}{b_0+b_1x+b_2x^2+b_3x^3},$$

where $a_0=2\beta^2(\gamma-1), a_1=2\beta(1-2\gamma), a_2=2\gamma-1,$

$b_0=0, b_1=2\beta^2, b_2=-3\beta, b_3=1$.

Conversely, if

$$f(x)=\frac{2\gamma}{\beta}\left(1-\frac{x}{\beta}\right)\left(\frac{x}{\beta}\left(2-\frac{x}{\beta}\right)\right)^{\gamma-1},$$

the probability density function (7.2) of the general form of the J-shaped distribution, then

$$\frac{f'(x)}{f(x)}=\frac{a_0+a_1x+a_2x^2}{b_0+b_1x+b_2x^2+b_3x^3},$$

where $a_0=2\beta^2(\gamma-1), a_1=2\beta(1-2\gamma), a_2=2\gamma-1$,

$b_0=0, b_1=2\beta^2, b_2=-3\beta, b_3=1$.

Proof.

We have

$$\frac{f'(x)}{f(x)}=\frac{a_0+a_1x+a_2x^2}{b_0+b_1x+b_2x^2+b_3x^3}$$

$$=\frac{2\beta^2(\gamma-1)+2\beta(1-2\gamma)x+(2\gamma-1)x^2}{2\beta^2x-3\beta x^2+x^3}$$

$$=\frac{-x(2\beta-x)+(\gamma-1)(\beta-x)(2\beta-x)-(\gamma-1)x(\beta-x)}{x(\beta-x)(2\beta-x)},$$

or, after simplifications, we have

$$\frac{f'(x)}{f(x)}=\frac{-1}{(\beta-x)}+\frac{(\gamma-1)}{x}-\frac{(\gamma-1)}{(2\beta-x)}. \tag{7.1.3}$$

Integrating both sides of the Equation (7.3) with respect to x, we obtain

$$\ln(f(x))=\ln(c)+\ln((\beta-x))+(\gamma-1)\ln(x)+(\gamma-1)\ln((2\beta-x))$$

or

$$f(x)=c(\beta-x)(2\beta x-x^2)^{(\gamma-1)}, \tag{7.1.4}$$

where c is a constant. Now, integrating both sides of the Equation (7.4) with respect to x, from $x=0$ to β, using the condition $\int_0^\beta f(x)=1$, and substituting $2\beta x-x^2=u$, we easily have $c=\frac{2\gamma}{\beta^{2\gamma}}$.

Substituting the above expression for the constant c in the Equation (7.1.4), we have

$$f(x)=\frac{2\gamma}{\beta^{2\gamma}}(\beta-x)(2\beta x-x^2)^{(\gamma-1)},$$

which can easily be expressed as

$$f(x)=\frac{2\gamma}{\beta}\left(1-\frac{x}{\beta}\right)\left(\frac{x}{\beta}\left(2-\frac{x}{\beta}\right)\right)^{\gamma-1},$$

the required probability density function (7.1,2) of the general form of the J-shaped distribution.

Conversely, let

$$f(x)=\frac{2\gamma}{\beta}\left(1-\frac{x}{\beta}\right)\left(\frac{x}{\beta}\left(2-\frac{x}{\beta}\right)\right)^{\gamma-1}, \tag{7.1.5}$$

which is the probability density function of the general form of the J-shaped distribution. By differentiating the Equation (7.1.5), with respect to x, we easily obtain the following function:

$$f'(x)=\frac{2\gamma}{\beta^2}\left(\frac{x}{\beta}\left(2-\frac{x}{\beta}\right)\right)^{\gamma-2}\left\{\left(1-\frac{x}{\beta}\right)^2(2\gamma-1)-1\right\}. \qquad (7.1.6)$$

Thus, from Equations (7.1.5) and (7.1.6), we have

$$\frac{f'(x)}{f(x)}=\frac{\frac{2\gamma}{\beta^2}\left(\frac{x}{\beta}\left(2-\frac{x}{\beta}\right)\right)^{\gamma-2}\left(\left(1-\frac{x}{\beta}\right)^2(2\gamma-1)-1\right)}{\frac{2\gamma}{\beta}\left(1-\frac{x}{\beta}\right)\left(\frac{x}{\beta}\left(2-\frac{x}{\beta}\right)\right)^{\gamma-1}}$$

$$=\frac{(\beta-x)^2(2\gamma-1)-\beta^2}{x(\beta-x)(2\beta-x)}$$

$$=\frac{2\beta^2(\gamma-1)+2\beta(1-2\gamma)x+(2\gamma-1)x^2}{2\beta^2x-3\beta x^2+x^3}$$

$$=\frac{a_0+a_1x+a_2x^2}{b_0+b_1x+b_2x^2+b_3x^3},$$

where $a_0=2\beta^2(\gamma-1), a_1=2\beta(1-2\gamma), a_2=2\gamma-1$,

$b_0=0, b_1=2\beta^2, b_2=-3\beta, b_3=1$.

This completes the proof of Theorem 7.1.

7.2. Characterizations

By considering a product of reverse hazard rate and another function of the truncated point, we shall now present some characterizations of the J-shaped distribution by the left and right truncated moment, by first considering the conditional expectations for $E(X)$, and, then, more in general, for $E(X^m)$, $m\geq 1$.. For details, see Ahsanullah (2017).

7.2.1. Truncated Moment Based on $E(X)$

In this Sub-Section, we will present the characterization results of the J-shaped.

Distribution by the left and right truncated conditional expectations for $E(X)$. For this, we shall need the following assumption:

Assumption 7.2.1: Suppose that X is an absolutely continuous (with respect to Lebesgue measure) random variable with *cdf* $F(x)$ and *pdf* $f(x)$. We assume $\beta = \inf\{x|F(x) > 0\}$, $\delta = \sup\{x|F(x) < 1\}$, $E(X)$ and $f^{/}(x)$ exist for all $x \in (\beta, \delta)$.

Case 1: We shall need the following lemma:

Lemma 7.2.1: Under the Assumption 7.2.1, if $E(X|X \le x) = g(x)\eta(x)$, where $g(x)$ is a differentiable function, and $\eta(x) = \frac{f(x)}{F(x)}$ for all $x > 0$, then we have

$$f(x) = ce^{\int_{\beta}^{x} \frac{u - g'(u)}{g(u)} du},$$

where c is determined by taking $\int_{\beta}^{\delta} f(x)dx = 1$.

Proof: We have

$$\frac{\int_{\beta}^{x} uf(u)du}{F(x)} = \frac{g(x)f(x)}{F(x)}.$$

Thus

$$\int_{\beta}^{x} uf(u)du = g(x)f(x).$$

Differentiating both sides of the above equation with respect to x, we obtain

$$xf(x) = g'(x)f(x) + g(x)f'(x).$$

On simplification, we get

$$\frac{f'(x)}{f(x)} = \frac{x - g'(x)}{g(x)} .$$

Integrating the above equation, we obtain

$$f(x) = ce^{\int_\beta^x \frac{u-g'(u)}{g(u)} du},$$

where c is determined such that $\int_\beta^\delta f(x)dx = 1.$ This completes the proof of Lemma 7.2.1.

Theorem 7.2.1 Without loss of generality, by assuming β = 1, in the general form of J-shaped distribution of the random variable X, its *pdf* and *cdf* are respectively given by

$$f(x) = 2\gamma(1-x)\left(x(2-x)\right)^{\gamma-1}, \quad 0 \le x \le 1, 0 < \gamma \le 1, \quad (7.2.1.1)$$

and

$$F(x) = \begin{cases} 0, & x < 0, \\ \left[x(2-x)\right]^\gamma, & 0 \le x \le 1, 0 < \gamma \le 1, \\ 1, & x > 1. \end{cases} \quad (7.2.1.2)$$

Suppose that the random variable X has an absolutely continuous (with respect to Lebesgue measure) cumulative distribution function (*cdf*) $F(x)$ and probability density function (*pdf*) $f(x)$, as given in (7.1.1)) and (7.1.2), respectively. Assume that $\gamma = 1$. We also assume that E(X) exists. Then X has a J-shaped distribution if and only if

$$E(X|X \le t) = g(t)\eta(t),$$

where

$$g(t) = \tfrac{t^2}{2} + \tfrac{t^3}{6(1-t)},$$

and

$$\eta(t) = \frac{f(t)}{F(t)}.$$

Proof:
Necessary Part: Suppose that

$$f(x) = 2(1-x), \qquad 0 \le x \le 1.$$

Then, since, from Lemma 7.2.1, we have

$$E(X|X \le t) = \frac{\int_0^t u\, f(u)\, du}{F(t)},$$

and

$$E(X|X \le t) = g(t)\eta(t) = g(t)\frac{f(t)}{F(t)},$$

we have

$$g(t) = \frac{\int_0^t u\, f(u)\, du}{f(t)}$$

$$= \frac{\int_0^t u\,[2(1-u)]\, du}{2(1-t)}$$

$$=\frac{3t^2-2t^3}{6(1-t)},$$

from which it is easily seen that

$$g(t)=\tfrac{t^2}{2}+\tfrac{t^3}{6(1-t)}. \qquad (7.2.1.3)$$

Consequently,

$$E(X|X\le t)=g(t)\eta(t),$$

where

$$\eta(t)=\frac{f(t)}{F(t)},$$

and, hence, the proof of "if" part of the Theorem 7.2.1 follows from Lemma 7.2.1.

Sufficiency Part: We will prove now the "only if" condition the Theorem 7.2.1. Suppose that

$$g(t)=\tfrac{t^2}{2}+\tfrac{t^3}{6(1-t)}.$$

Then, after simple differentiation and simplification, we have

$$g'(t)=t+\tfrac{1}{1-t}\left(\tfrac{t^2}{2}-\tfrac{t^3}{6(1-t)}\right)$$

Thus

$$\frac{f'(t)}{f(t)}=\tfrac{t-g'(t)}{g(t)}=-\tfrac{1}{1-t}.$$

On integrating the above equation, we have

$$f(x) = ce^{-\int_0^x \frac{1}{1-t}dt} = c(1-x),$$

where c is a constant.

Using the boundary condition

$$\int_0^1 f(x)dx = 1,$$

we obtain

$$f(x) = 2(1-x), \qquad 0 \le x \le 1.$$

This completes the proof of Theorem 7.2.1.

Theorem 7.2.2: Suppose that the random variable X has an absolutely continuous (with respect to Lebesgue measure) cumulative distribution function (cdf) $F(x)$ and probability density function (pdf) $f(x)$, as given in (7.7) and (7.8), respectively, with $\gamma > 1$. We assume that $E(X)$ and $f'(x)$ exist. Then X has a J-shaped distribution if and only if

$$E(X \mid X \le t) = g(t)\eta(t),$$

where $g(t) = \frac{t^2(2-t)}{2\gamma(1-t)} - \frac{\int_0^t (x(2-x))^{\gamma} dx}{2\gamma(1-t)(t(2-t))^{\gamma-1}}$,

and $\eta(t) = \dfrac{f(t)}{F(t)}$.

Proof:

Necessary Part: Suppose that

$$f(x) = 2\gamma(1-x)\left(x(2-x)\right)^{\gamma-1}, \qquad 0 \le x \le 1,\, 0 < \gamma < 1.$$

Then it is easily seen that

$$g(t) = \frac{t^2(2-t)}{2\gamma(1-t)} - \frac{\int_0^t (x(2-x))^\gamma \, dx}{2\gamma(1-t)(t(2-t))^{\gamma-1}} .$$

Consequently,

$$E(X \mid X \le t) = g(t)\eta(t),$$

where

$$\eta(t) = \frac{f(t)}{F(t)},$$

and, hence, the proof of "if" part of the Theorem 7.2.2 follows easily from Lemma 7.2.1.

Sufficiency Part: We will prove now the "only if" condition the Theorem 7.2.2. Suppose that

$$g(t) = \frac{t^2(2-t)}{2\gamma(1-t)} - \frac{\int_0^t (x(2-x))^\gamma \, dx}{2\gamma(1-t)(t(2-t))^{\gamma-1}} .$$

Then, after simple differentiation and simplification, we have

$$\begin{aligned}
g'(t) &= \frac{t^2(2-t)}{2\gamma(1-t)}\left[\frac{1}{t} - \frac{1}{2-t} + \frac{1}{1-t}\right] + \frac{\int_0^t (x(2-x))^\gamma \, dx}{2\gamma(1-t)(t(2-t))^{\gamma-1}}\left(\frac{-1}{1-t} - (\gamma-1)\left(\frac{1}{t} - \frac{1}{2-t}\right)\right) \\
&= \frac{t^2(2-t)}{2\gamma(1-t)}\left[\frac{1}{t} - \frac{1}{2-t} + \frac{1}{1-t}\right] + (g(t) - (g(t) - \frac{t^2(2-t)}{2\gamma(1-t)})\left(\frac{-1}{1-t} - (\gamma-1)\left(\frac{1}{t} - \frac{1}{2-t}\right)\right) \\
&= g(t)\left(\frac{-1}{1-t} - (\gamma-1)\left(\frac{1}{t} - \frac{1}{2-t}\right)\right) + \frac{t^2(2-t)}{2\gamma(1-t)}\gamma\left(\frac{1}{t} - \frac{1}{2-t}\right) \\
&= g(t)\left(\frac{-1}{1-t} - (\gamma-1)\left(\frac{1}{t} - \frac{1}{2-t}\right)\right) + t
\end{aligned}$$

Thus

$$\frac{t - g'(t)}{g(t)} = -\frac{-1}{1-t} + (\gamma-1)\left(\frac{1}{t} - \frac{1}{2-t}\right).$$

Hence, on integrating the above equation, we have

$$f(x) = ce^{\int_0^x (-\frac{-1}{1-t} + (\gamma-1)(\frac{1}{t} - \frac{1}{2-t}))dt}$$
$$= c(1-x)(x(2-x))^{\gamma-1},$$

where c is a constant.

Using the boundary condition

$$\int_0^1 f(x)dx = 1,$$

we obtain

$$f(x) = 2\gamma(1-x)(x(2-x))^{\gamma-1}, \quad 0 \le x \le 1, 0 < \gamma < 1.$$

This completes the proof of Theorem 7.2.2.

Case 2: We shall need the following lemma:

Lemma 7.2.2: Under the Assumption 7.2.1, if $E(X \mid X \ge x) = h(x)r(x)$, where $r(x) = \frac{f(x)}{1-F(x)}$ and $h(x)$ is a continuous differentiable function of x with the condition that $\int_x^\infty \frac{u + [h(u)]'}{h(u)} du$ is finite for $x > 0$, then $f(x) = ce^{-\int_0^x \frac{u + [h(u)]'}{h(u)} du}$, where c is a constant determined by the condition $\int_0^\infty f(x)dx = 1$.

Proof. Suppose that $E(X \mid X \ge x) = h(x)r(x)$. Then, since

$E(X \mid X \ge x) = \frac{\int_x^\infty u f(u)du}{1-F(x)}$ and $r(x) = \frac{f(x)}{1-F(x)}$, we have

$h(x) = \frac{\int_x^\infty u f(u)du}{f(x)}$, that is, $\int_x^\infty u f(u)du = f(x)h(x)$.

Differentiating the above equation with respect to respect to x, we obtain

$$-x f(x) = f'(x) h(x) + f(x) [h(x)]'.$$

From the above equation, we obtain.

$$\frac{f'(x)}{f(x)} = -\frac{x + [h(x)]'}{h(x)}$$

On integrating the above equation with respect to x, we have

$$f(x) = c e^{-\int_0^x \frac{u + [h(u)]'}{h(u)} du},$$

where c is obtained by the condition $\int_0^\infty f(x) dx = 1$. This completes the proof of Lemma 7.2.2.

Theorem 7.2.3. If the random variable X satisfies the Assumption 7.2.1 with $\beta = 0$ and $\delta = 1$, then $E(X \mid X \geq x) = h(x) r(x)$, where $r(x) = \frac{f(x)}{1 - F(x)}$, and

$$h(x) = \frac{\int_x^1 u f(u) du}{f(x)} = \frac{E(X)}{f(x)} + \frac{\int_0^x u f(u) du}{f(x)} = \frac{E(X)}{f(x)} + \frac{g(x)}{f(x)},$$

if and only if X has the distribution with the *pdf* (1), where $g(x)$ is given by (7.2.1.3), and $f(x)$ represents the *pdf* (7.2.\2), with $\gamma = 1$, and $E(X)$, the first moment, is given as : $E(X) = 1 - \frac{\Gamma(\frac{3}{2})\Gamma(2)}{\Gamma(\frac{5}{2})}$.

Proof. Suppose that $E(X|X \geq x) = h(x)\frac{f(x)}{1-F(x)}$. Then, since $E(X|X \geq x) = \frac{\int_x^1 u\, f(u)\, du}{1-F(x)}$, we have $h(x) = \frac{\int_x^1 u\, f(u)\, du}{f(x)}$. Now, if the random variable X satisfies the Assumption 7.2.1 and has the distribution with the *pdf* as given in (7.1.2) with $\gamma = 1$, then we have

$$h(x) = \frac{\int_x^1 u\, f(u)\, du}{f(x)} = \frac{E(X)}{f(x)} + \frac{\int_0^x u\, f(u)\, du}{f(x)} = \frac{E(X)}{f(x)} + \frac{g(x)}{f(x)},$$

where $g(x)$ is given by (7.9), and $f(x)$ represents the *pdf* (7.1.2), with $\gamma = 1$, and $E(X)$, the first moment, is given as: $E(X) = 1 - \frac{\Gamma(\frac{3}{2})\Gamma(2)}{\Gamma(\frac{5}{2})}$.

Conversely, suppose that

$$h(x) = \frac{E(X)}{f(x)} + \frac{g(x)}{f(x)},$$

where $g(x)$ is given by (7.2.1.3) and $f(x)$ represents the *pdf* (7.7), with $\gamma = 1$, and $E(X)$, the first moment, is given by Equation (4.1.4.), Chapter 3, as follows: $E(X) = 1 - \frac{\Gamma(\frac{3}{2})\Gamma(2)}{\Gamma(\frac{5}{2})}$.

Then, differentiating both sides of the equation $h(x) = \frac{\int_x^1 u\, f(u)\, du}{f(x)}$, with respect to x, using Lemma 7.2.2, and simplifying, we have

$$(h(x))' = -x - h(x)\left(\frac{-1}{1-x}\right),$$

from which we obtain

$$\frac{x+\left(h(x)\right)'}{h(x)}=\frac{1}{1-x}.$$

Since, by Lemma 7.2.2, we have

$$\frac{f'(x)}{f(x)}=-\frac{x+\left[h(x)\right]'}{h(x)}$$

it follows that

$$\frac{f'(x)}{f(x)}=-\frac{1}{1-x}.$$

On integrating the above expression with respect to x and simplifying, we obtain

$$\ln f(x)=\ln\left(c(1-x)\right),$$

or,

$$f(x)=c(1-x),$$

where c is the normalizing constant to be determined. Thus, on integrating the above equation with respect to x from $x=0$ to $x=1$, using the condition $\int_0^1 f(x)dx=1$, we easily obtain

$$c=2,$$

and, hence, we have

$$f(x)=2(1-x), \qquad 0\le x\le 1,$$

which is the required *pdf* (7.7), with $\gamma = 1$. This completes the proof of Theorem 7.2.3.

Remark: If the random variable X has an absolutely continuous (with respect to Lebesgue measure) cumulative distribution function (*cdf*) $F(x)$ and probability density function (*pdf*) $f(x)$, as given in (7.1.1) and (7..1.2), respectively, with $\gamma > 1$, then the proof is similar to

Theorem 7.2.2.

7.2.2. Order Statistics

If $X_1, X_2, \ldots, X_n$ be the n independent copies of the random variable X with absolutely continuous distribution function $F(x)$ and *pdf* $f(x)$, and if $X_{1,n} \le X_{2,n} \le \ldots \le X_{n,n}$ be the corresponding order statistics, then it is known from Arnold et al. (2005), chapter 2, or Ahsanullah et al. (2013), chapter 5, that $X_{j,n} \mid X_{k,n} = x$, for $1 \le k < j \le n$, is distributed as the $(j-k)th$ order statistics from $(n-k)$ independent observations from the random variable V having the *pdf* $f_V(v \mid x)$ where $f_V(v \mid x) = \dfrac{f(v)}{1-F(x)}$, $0 \le v < x$, and $X_{i,n} \mid X_{k,n} = x, 1 \le i < k \le n$, is distributed as *ith* order statistics from k independent observations from the random variable W having the *pdf* $f_W(w \mid x)$ where $f_W(w \mid x) = \dfrac{f(w)}{F(x)}$, $w < x$.

Let $S_{k-1} = \frac{1}{k-1}\left(X_{1,n} + X_{2,n} + \ldots + X_{k-1,n}\right)$,

and $T_{k,n} = \frac{1}{n-k}\left(X_{k+1,n} + X_{k+2,n} + \ldots + X_{n.n}\right)$.

Theorem 7.2.4. If the random variable X satisfies the Assumption 7.2.1 with $\beta = 0$ and $\delta = 1$, then $E(S_{k-1} \mid X_{k,n} = x) = g(x)\eta(x)$, where $\eta(x) = g(t)\dfrac{f(t)}{F(t)}$, and $g(x)$ being given by the equation,

$g(x)=\frac{x^2}{2}+\frac{x^3}{6(1-x)}$, (see Theorem 7.2.1), if and only if X has a J-shaped distribution with the *pdf* given by

$$f(x)=2(1-x), \qquad 0\le x\le 1.$$

Proof: It is known that $E(S_{k-1} \mid X_{k,n}=x)=E(X \mid X\le x)$; see David and Nagaraja (2003), an Ahsanullah et al. (2013). Hence, by Theorem 7.2.1, the result follows.

Theorem 7.2.5. If the random variable X satisfies the Assumption 7.2.1 with $\beta=0$ and $\delta=1$,

then $E(T_{k,n} \mid X_{k,n}=x)=r(x)\dfrac{f(x)}{1-F(x)}$, where

$r(x)=\dfrac{\left(E(X)-g(x)f(x)\right)}{f(x)}$, $g(x)$ given by the equation,

$g(x)=\frac{x^2}{2}+\frac{x^3}{6(1-x)}$, (see Theorem 7.2.1), and $E(X)$, the first moment, is given by the Equation (4.17), $E(X)=1-\frac{\Gamma(\frac{3}{2})\Gamma(2)}{\Gamma(\frac{5}{2})}$, (see Chapter 3), if and only if X has a J-shaped distribution with the *pdf* given by $f(x)=2(1-x), \quad 0\le x\le 1$.

Proof: Since $E(T_{k,n} \mid X_{k,n}=x)=E(X \mid X\ge x)$, David and Nagaraja (2003), and Ahsanullah et al. (2013), the result follows from Theorem 7.2.3.

7.2.3. Upper Record Values

For details on record values, see Ahsanullah (1995). Let $X_1, X_2, \ldots$ be a sequence of independent and identically distributed absolutely continuous random variables with distribution function $F(x)$ and *pdf* $f(x)$. If $Y_n=\max(X_1, X_2, \ldots, X_n)$ for $n\ge 1$ and $Y_j>Y_{j-1}, j>1$, then X_j is called an upper record value of $\{X_n, n\ge 1\}$. The indices at which the upper records occur are given by the record times $\{U(n) > \min(j \mid j > U(n+1), X_j > X_{U(n-1)}, n > 1)\}$ and $U(1) = 1$. Let the nth upper record value be denoted by $X(n) = X_{U(n)}$.

Theorem 7.2.6. If the random variable X satisfies the Assumption 7.2.1 with $\beta = 0$ and

$\delta = 1$, then $E(X(n+1) \mid X(n) = x) = r(x)\frac{f(x)}{1-F(x)}$, where

$r(x) = \frac{(E(X) - g(x)f(x))}{f(x)}$, $g(x)$ given by the equation, $g(x) = \frac{x^2}{2} + \frac{x^3}{6(1-x)}$,

and $E(X)$, the first moment, is given as $E(X) = 1 - \frac{\Gamma(\frac{3}{2})\Gamma(2)}{\Gamma(\frac{5}{2})}$, (see Chapter 3), if and only if X has a J shaped distribution with the *pdf* given by $f(x) = 2(1-x), \quad 0 \le x \le 1$.

Proof: It is known from David and Nagaraja (2003), and Ahsanullah et al. (2013) that $E(X(n+1) \mid X(n) = x) = E(X \mid X \ge x)$. Then, the result follows from Theorem 7.2.3.

7.2.4. Truncated Moment Based on $E(X^m)$, $m \ge 1$

In this Sub-Section, we will present the characterization results of the J-shaped distribution by the left and right truncated conditional expectations for $E(X^m)$, where $m \ge 1$ is a positive integer. We shall need the following assumption and lemmas:

Assumption 7.2.2: Suppose that X is an absolutely continuous (with respect to Lebesgue measure) random variable with *cdf* $F(x)$ and *pdf* $f(x)$. We assume that $\beta = \inf\{x \mid F(x) > 0\}$, and $\delta = \sup\{x \mid F(x) < 1\}$. We assume that $E(X^m)$, $m \ge 1$, and $f'(x)$ exist for all $x \in (\beta, \delta)$.

Lemma 7.2.3: Under the Assumption 7.2.2, if $E(X^m \mid X \le x) = g(x)\eta(x)$, where $g(x)$ is a differentiable function, and $\eta(x) = \frac{f(x)}{F(x)}$ for all $x > 0$, then we have $f(x) = ce^{\int \frac{x^m - g'(x)}{g(x)} dx}$ where c is a constant, determined by taking $\int_\beta^\delta f(x)dx = 1$.

Proof. We have

$$\frac{\int_{\beta}^{x} u^m f(u)\,du}{F(x)} = \frac{g(x)f(x)}{F(x)},$$

or,

$$\int_{\beta}^{x} u^m f(u)\,du = g(x)f(x).$$

Differentiating both sides of the above equation with respect to x, we obtain

$$x^m f(x) = g'(x)f(x) + g(x)f'(x).$$

On simplification, we get

$$\frac{f'(x)}{f(x)} = \frac{x^m - g'(x)}{g(x)}.$$

Integrating the above equation, we obtain

$$f(x) = c\,e^{\int \frac{x^m - g'(x)}{g(x)}dx}$$

where c is a constant, determined by taking $\int_{\beta}^{\delta} f(x)dx = 1$.

Lemma 7.2.4: Suppose that X is an absolutely continuous (with respect to Lebesgue measure) random variable with *cdf* $F(x)$ and *pdf* $f(x)$. We assume that $\beta = \inf\{x|F(x) > 0\}$, and $\delta = \sup\{x|F(x) < 1\}$. We assume that $E\left(X^m\right)$, $m \geq 1$, and $f'(x)$ exist for all $x \in (\beta, \delta)$. if $E(X^m|X \geq x) = h(x)r(x)$, where h(x) is a continuous differentiable function in

$\beta \le x \le \delta$ and $r(x)=\frac{f(x)}{1-F(x)}$, if and only if f(x) = c exp(- $\int \frac{x^m + h^{/}(x)}{h(x)}$ dx),

where c is determined by the condition $\int_{\beta}^{\delta} f(x)dx$ = 1.

Proof. The proof is similar to Lemma 7.2.3.

We assume that $\beta = 0 \; and \; \delta = 1$. Then, if the random variable X has the general form of J-shaped distribution, its *pdf* $f(x)$ and *cdf* $F(x)$, are respectively given by (7.7) and (7.8).

Theorem 7.2.7. Under the Assumption 7.2.2, suppose that $E(X^n)$, $n \ge 1$, exists. Then

$$E\left(X^n \mid X \le x\right) = g(x)\eta(x),\ \eta(x) = \frac{f(x)}{F(x)},$$

where

$$g(x) = \frac{\sum_{j=0}^{n} (-1)^{n-j} \binom{n}{j} \left[B\left(\frac{n-j+2}{2}, \gamma\right) - B_{(1-x)^2}\left(\frac{n-j+2}{2}, \gamma\right) \right]}{2(1-x)\left(x(2-x)\right)^{\gamma-1}}, \tag{7.2.1.4}$$

if and only if X has a J-shaped distribution with the *pdf* given by

$$f(x) = 2\gamma(1-x)\left(x(2-x)\right)^{\gamma-1}, 0 \le x \le 1, 0 < \gamma \le 1.$$

Proof: Suppose that

$$f(x) = 2\gamma(1-x)\left(x(2-x)\right)^{\gamma-1}, 0 \le x \le 1, 0 < \gamma \le 1.$$

Then, since, from Lemma 7.2.3, we have

$$E\left(X^n \middle| X \le x\right) = \frac{\int_0^x u^n f(u)du}{F(x)},$$

and

$$E\left(X^n \mid X \le x\right) = g\left(x\right)\eta(x) = g\left(x\right)\frac{f(x)}{F(x)},$$

we have

$$g\left(x\right) = \frac{\int_0^x u^n f\left(u\right)du}{f\left(x\right)} = \frac{I_X\left(x\right)}{f\left(x\right)}, \tag{7.2.1.6}$$

where the integral in the numerator, $I_X\left(x\right) = \int_0^x u^n f\left(u\right)du$, denotes the incomplete moment of X^n, and is determined as follows:

$$I_X\left(x\right) = \int_0^x u^n f\left(u\right)du = \int_0^x u^n\, 2\gamma\left(1-u\right)\left(u\left(2-u\right)\right)^{\gamma-1} du\,. \tag{7.2.1.7}$$

Letting $u\left(2-u\right)=t$, and subsequently, $1-t=z$, in the above integral (7.12), after simplifications, we have

$$\begin{aligned}
I_X\left(x\right) &= \left(\gamma\right)\int_{(1-x)^2}^1 \left(1-z^{1/2}\right)^n \left(1-z\right)^{\gamma-1} dz \\
&= \left(\gamma\right)\sum_{j=0}^n \left(-1\right)^{n-j}\binom{n}{j}\int_{(1-x)^2}^1 \left(z\right)^{\frac{(n-j)}{2}} \left(1-z\right)^{\gamma-1} dz \\
&= \left(\gamma\right)\sum_{j=0}^n \left(-1\right)^{n-j}\binom{n}{j}\left[\int_0^1 \left(z\right)^{\frac{(n-j)}{2}} \left(1-z\right)^{\gamma-1} dz - \int_0^{(1-x)^2} \left(z\right)^{\frac{(n-j)}{2}} \left(1-z\right)^{\gamma-1} dz\right] \\
&= \left(\gamma\right)\sum_{j=0}^n \left(-1\right)^{n-j}\binom{n}{j}\left[B\left(\frac{n-j+2}{2},\gamma\right) - B_{(1-x)^2}\left(\frac{n-j+2}{2},\gamma\right)\right],
\end{aligned} \tag{7.2.1.8}$$

where $B\left(.,.\right)$ and $B_{(1-x)^2}\left(.,.\right)$ denote the beta and the incomplete beta functions, respectively, defined as follows:

$$B(a,b)=\int_0^1 t^{a-1}(1-t)^{b-1}\,dt=\frac{\Gamma(a)\Gamma(b)}{\Gamma(a+b)}=B(b,a),$$

and

$$B_v(p,q)=\int_0^v t^{p-1}(1-t)^{q-1}\,dt=\frac{v^p}{p}\,{}_2F_1(p,1-q;p+1;v),$$

where ${}_2F_1(p,1-q;p+1;v)$ denotes the generalized hypergeometric function ${}_2F_1$ of order (2, 1) or the Gauss hypergeometric function; cf.8.380.1, page 948, 8.391, page 950, and 9.1, page 1039, Gradshteyn and Ryzhik (2000), and 6.6.1, page 263, and 15.1 page 556, Abramowitz and Stegun (1972).

Using the expression (7.13) for $I_X(x)$ and the expression for the *pdf* of the J-shaped distribution in the Equation (7.11), we have

$$g(x)=\frac{\sum_{j-0}^{n}(-1)^{n-j}\binom{n}{j}\left[B\left(\frac{n-j+2}{2},\gamma\right)-B_{(1-x)^2}\left(\frac{n-j+2}{2},\gamma\right)\right]}{2(1-x)\left(x(2-x)\right)^{\gamma-1}}$$

Consequently, the proof of "if" part of the Theorem 7.2.4 follows from Lemma 7.2.3.

Now, for the sufficient condition, suppose that

$$g(x)=\frac{\sum_{j-0}^{n}(-1)^{n-j}\binom{n}{j}\left[B\left(\frac{n-j+2}{2},\gamma\right)-B_{(1-x)^2}\left(\frac{n-j+2}{2},\gamma\right)\right]}{2(1-x)\left(x(2-x)\right)^{\gamma-1}}$$

As above, we have

$$g(x)=\frac{\int_0^x u^n f(u)\,du}{f(x)},$$

or

$$\int_0^x u^n f(u)du = f(x)g(x).$$

Differentiating the above equation with respect to respect to x, we obtain

$$x^n f(x) = f'(x)g(x) + f(x)g'(x),$$

from which, using the definition of the *pdf* $f(x)$ of the J-shaped distribution and its derivative $f'(x)$, we easily obtain

$$g'(x) = x^n - g(x)\frac{\left[-(x(2-x))^{\gamma-1} + (\gamma-1)(1-x)(x(2-x))^{\gamma-2}(2-2x)\right]}{(1-x)\,(x(2-x))^{\gamma-1}},$$

or,

$$\frac{x^n - g'(x)}{g(x)} = -\frac{1}{(1-x)} + \frac{(\gamma-1)}{x(2-x)}$$

$$= -\frac{1}{(1-x)} + \frac{(\gamma-1)}{x} - \frac{(\gamma-1)}{(2-x)}. \qquad (7.2.1.9)$$

But, from Lemma 7.2.3, we have

$$\frac{f'(x)}{f(x)} = \frac{x^n - g'(x)}{g(x)}. \qquad (7.2.1.10)$$

Therefore, from (7.14) and (7.15), it follows that

$$\frac{f'(x)}{f(x)} = -\frac{1}{(1-x)} + \frac{(\gamma-1)}{x} - \frac{(\gamma-1)}{(2-x)}. \qquad (7.2.1.11)$$

Now, integrating Equation (7.2,1.11) with respect to x and simplifying, we easily have

$$\ln\left(f(x)\right)=\ln\left(c(1-x)\left[x(2-x)\right]^{\gamma-1}\right),$$

or

$$f(x)=c(1-x)\left[x(2-x)\right]^{\gamma-1}, \tag{7.2.1.12}$$

where c is the normalizing constant. Thus, on integrating the above equation (7.17) with respect to x from $x=0$ to $x=1$, and using the condition $\int_0^1 f(x)dx=1$, we easily obtain $c=2\gamma$.

Thus we have

$$f(x)=2\gamma(1-x)\left(x(2-x)\right)^{\gamma-1}, 0\le x\le 1, 0<\gamma\le 1.$$

This completes the proof of Theorem 7.2.7.

Theorem 7.2.8: Under the Assumption 7.2.2, suppose that $E\left(X^n\right)$, $n\ge 1$, exists. Then $E\left(X^n \mid X\ge x\right)=r(x)\frac{f(x)}{1-F(x)}$, where $r(x)=\frac{\left(E\left(X^n\right)-g(x)f(x)\right)}{f(x)}$, $g(x)$ being given by the Equation (7.10), and $E\left(X^n\right)$, the 1st moment of X^n, given by $E(X^n)=1-\sum_{j=1}^{n}\binom{n}{j}(-1)^j\frac{\Gamma\left(\frac{j+2}{2}\right)\Gamma(\gamma+1)}{\Gamma\left(\frac{j+2}{2}+\gamma\right)}$, see the Equation (4.16), Chapter 3, if and only if X has a J-shaped distribution with the *pdf* given by

$$f(x)=2\gamma(1-x)\left(x(2-x)\right)^{\gamma-1}, 0\le x\le 1, 0<\gamma\le 1.$$

Proof: Suppose that $E\left(X^n \mid X \geq x\right) = r(x)\frac{f(x)}{1-F(x)}$. Then, since $E\left(X^n \mid X \geq x\right) = \frac{\int_x^1 u^n f(u)du}{1-F(x)}$, proceeding in the same way as in Theorem 7.2.7, following the similar arguments, and using the result of Lemma 7.2.4, we have

$$r(x) = \frac{\int_x^1 u^n f(u)du}{f(x)}$$
$$= \frac{\int_0^1 u^n f(u)du - \int_0^x u^n f(u)du}{f(x)}$$
$$= \frac{\left(E\left(X^n\right) - g(x) f(x)\right)}{f(x)},$$

where $f(x)$ denotes the *pdf* of the J-shaped distribution given by

$$f(x) = 2\gamma(1-x)\left(x(2-x)\right)^{\gamma-1}, 0 \leq x \leq 1, 0 < \gamma \leq 1,$$

$g(x)$ being given by the Equation (7.10), and $E\left(X^n\right)$, the 1st moment of X^n, as given above. Consequently, the proof of "if" part of the Theorem 7.2.8 follows.

Conversely, for the sufficiency part, suppose that $r(x) = \frac{\left(E\left(X^n\right) - g(x) f(x)\right)}{f(x)}$. Now, using the result of Lemma 7.2.4, we have

$$r(x) = \frac{\int_x^1 u^n f(u)du}{f(x)},$$

or

$$\int_x^1 u^n f(u)du = f(x).r(x).$$

Differentiating the above equation with respect to respect to x, we obtain

$$-x^n f(x) = f'(x).r(x) + f(x).(r(x))'.$$

Thus, proceeding in the same way as in Theorem 7.2.7, and following the similar arguments, we easily obtain

$$f(x) = c(1-x)\left[x(2-x)\right]^{\gamma-1},$$

where the normalizing constant is given by

$$c = 2\gamma.$$

Thus, we have

$$f(x) = 2\gamma(1-x)\left(x(2-x)\right)^{\gamma-1}, 0 \le x \le 1, 0 < \gamma \le 1.$$

This completes the proof of Theorem 7.2.8.

7.3. Conclusion

Characterizations of probability distributions play important roles in the study of lifetime distributions. As pointed out by Glänzel (1990), a characterization of a particular probability distribution states that it is the only distribution that satisfies some specified conditions. As such, motivated by the importance of the practicality of the characterizations of lifetime distribution, in this chapter, we have provided the characterizations of the distribution of the J-shaped distribution, by truncated moment, by considering a product of reverse hazard rate and another function of the truncated point. We have also investigated the characterizations of the J-shaped distribution by order statistics and upper record values. We hope that the findings of this chapter will be very useful for the practitioners in various fields of studies and further enhancement of research in distribution theory, and its applications.

Chapter 8

Percentile Points

8. Introduction

The shapes of probability distributions in many fields of research, such as, biology, computer science, control theory, economics, engineering, genetics, hydrology, medicine, number theory, statistics, physics, psychology, reliability, risk management, etc., exhibit J-shaped distributions. Therefore, before any statistical applications of the J-shaped distribution, it would also be important to know its percentile points. The computations of the percentile points of the J-shaped distribution would be very useful for the practitioners in the above-said fields of studies and further enhancement of research of the J-shaped distribution, and its applications. For example, we may be interested in knowing the median (50%), 25%, or 75% quartiles. Similarly, it is necessary to compute the 90%, 95%, or 99% confidence levels for other applications in order to assess the statistical significance of an observation whose distribution is known. Thus, motivated by these facts, in this chapter, we have computed the percentage points of the J-shaped distribution of a random variable X.

8.1. Percentile of Order p

For any p, where $0 < p < 1$, the $(100\,p)th$ percentile or the quantile of order p of a probability distribution with the *pdf* $f_X(x)$ is a number x_p such that the area under $f_X(x)$ to the left of x_p is p. That is, the percentile points x_p are any roots of the equation

$$F(x_p) = \int_{-\infty}^{x_p} f_X(u)\,du = p\,,$$

or,

$$x_p = F^{-1}(p),$$

where $F(x_p)$ denotes the *cdf* of the above-said probability distribution. Thus, the percentile points x_p of the above-said probability distribution can be computed by numerically solving the equations for the *cdf* $F(x_p)$, that is, $x_p = F^{-1}(p)$, for any p, where $0 < p < 1$, by taking different sets of values of the parameters.

8.2. Percentile Points

For this, we consider the general form of the J-shaped distribution of the random variable X with the *pdf* given by

$$f(x) = \frac{2\gamma}{b}\left(1 - \frac{x}{b}\right)\left(\frac{x}{b}\left(2 - \frac{x}{b}\right)\right)^{\gamma - 1}, \quad 0 \le x \le b < \infty, 0 < \gamma \le 1, \quad (8.2.1)$$

and the corresponding *cdf* is given by

$$F(x) = \begin{cases} 0, & x < 0, \\ \left[\frac{x}{b}\left(2 - \frac{x}{b}\right)\right]^{\gamma}, & 0 \le x \le b < \infty, 0 < \gamma \le 1, \\ 1, & x > b, \end{cases} \quad (8.2.2)$$

The percentile points x_p associated with the *cdf* (8.2) of the J-shaped distribution of the random variable X is given by

$$F(x_p) = \frac{2\gamma}{b} \int_0^{x_p} \left(1 - \frac{t}{b}\right)\left(\frac{t}{b}\left(2 - \frac{t}{b}\right)\right)^{\gamma - 1} dt = p,$$

$$0 \le x_p \le b < \infty, 0 < \gamma \le 1. \tag{8.2.3}$$

Thus, using a Maple software program, we have numerically solved the equation

$$F(x_p) = \frac{2\gamma}{b} \int_0^{x_p} \left(1 - \frac{t}{b}\right)\left(\frac{t}{b}\left(2 - \frac{t}{b}\right)\right)^{\gamma - 1} dt = p, \tag{8.2.4}$$

that is,

$$x_p = F^{-1}(p), \tag{8.2.5}$$

and computed the percentage points x_p of the J-shaped distribution of the random variable X with the *pdf* (8.1), associated with the *cdf* (8.2), $F(x_p)$, for different sets of values of the parameters, $\gamma = 0.1, 0.2, 0.3, 0.4, 0.5, 0.6, 0.7, 0.8, 0.9$, when b = 0.2, 0.5, 1, respectively, which are provided in Tables 8.2.1–8.2.3.

Remark 8.2.1: Using the above Table 8.2.1, and the equation (8.1.4), by taking some values of the parameters, for example, $\gamma = 0.1, 0.9$, when b = 0.2, it can easily be verified that

$$F(0.0057129) = \left[\frac{2(0.1)}{0.2}\right] \int_0^{0.0057129} \left(1 - \frac{t}{0.2}\right)\left(\frac{t}{0.2}\left(2 - \frac{t}{0.2}\right)\right)^{0.1 - 1} dt = 0.75,$$

and

$$F(0.1789240) = \left[\frac{2(0.9)}{0.2}\right] \int_0^{0.1789240} \left(1 - \frac{t}{0.2}\right)\left(\frac{t}{0.2}\left(2 - \frac{t}{0.2}\right)\right)^{0.9 - 1} dt = 0.99$$

Similarly, it can be verified for other values of the parameters.

Table 8.2.1. Percentiles of the J-shaped Distribution of the random variable X with the pdf (8.2.1), when b = 0.2, and $\gamma = 0.1, 0.2, 0.3, 0.4, 0.5, 0.6, 0.7, 0.8, 0.9$

b	γ	75 %	80 %	85 %	90 %	95 %	99 %
0.2	0.1	0.0057129	0.0110422	0.0207655	0.0385910	0.0733093	0.1381557
	0.2	0.0253351	0.0360098	0.0508297	0.0720141	0.1048750	0.1557236
	0.3	0.0429394	0.0551275	0.0706541	0.0911592	0.1207133	0.1636978
	0.4	0.0567714	0.0692228	0.0844339	0.1037573	0.1306165	0.1684958
	0.5	0.0677124	0.0800000	0.0946435	0.1128220	0.1375500	0.1717865
	0.6	0.0765676	0.0885404	0.1025771	0.1197386	0.1427509	0.1742232
	0.7	0.0838973	0.0955085	0.1089641	0.1252377	0.1468377	0.1761211
	0.8	0.0900825	0.1013275	0.1142461	0.1297443	0.1501585	0.1776533
	0.9	0.0953875	0.1062789	0.1187076	0.1335246	0.1529261	0.1789240

Table 8.2.2. Percentiles of the J-shaped Distribution of the random variable X with the pdf (8.2.1), when b = 0.5, and $\gamma = 0.1, 0.2, 0.3, 0.4, 0.5, 0.6, 0.7, 0.8, 0.9$

b	γ	75 %	80 %	85 %	90 %	95 %	99 %
0.5	0.1	0.0142824	0.0276056	0.0519136	0.0964775	0.1832734	0.3453893
	0.2	0.0633379	0.0900244	0.1270742	0.1800352	0.2621875	0.3893090
	0.3	0.1073484	0.1378187	0.1766353	0.2278979	0.3017832	0.4092446
	0.4	0.1419285	0.1730571	0.2110848	0.2593932	0.3265412	0.4212396
	0.5	0.1692811	0.2000000	0.2366087	0.2820551	0.3438751	0.4294663
	0.6	0.1914191	0.2213511	0.2564428	0.2993466	0.3568772	0.4355581
	0.7	0.2097432	0.2387712	0.2724101	0.3130941	0.3670942	0.4403028
	0.8	0.2252063	0.2533187	0.2856154	0.3243607	0.3753962	0.4441333
	0.9	0.2384687	0.2656973	0.2967691	0.3338114	0.3823152	0.4473101

Remark 8.2.2: Similar to the Remark 8.2.1, using the above Table 8.2.2, the computed percentile points can easily be verified from the Equation (8.2.4) for the given values of the parameters.

Remark 8.2.3: Similar to the Remark 8.2.1, using the above Table 8.2.3, the computed percentile points can easily be verified from the Equation (8.3.4) for the given values of the parameters.

Table 8.2.3. Percentiles of the J-shaped Distribution of the random variable X with the *pdf* (8.2.1), and $\gamma = 0.1, 0.2, 0.3, 0.4, 0.5, 0.6, 0.7, 0.8, 0.9$

b	γ	75%	80%	85%	90%	95%	99%
1	0.1	0.0285647	0.0552112	0.1038273	0.1929550	0.3665467	0.6907785
	0.2	0.1266757	0.1800488	0.2541483	0.3600703	0.5243800	0.7786181
	0.3	0.2146968	0.2756373	0.3532706	0.4557958	0.6035664	0.8184892
	0.4	0.2838571	0.3461142	0.4221697	0.5187864	0.6530824	0.8424792
	0.5	0.3385622	0.4000000	0.4732173	0.5641101	0.6877501	0.8589326
	0.6	0.3828382	0.4427021	0.5128856	0.5986931	0.7137543	0.8711161
	0.7	0.4194864	0.4775424	0.5448203	0.6261883	0.7341884	0.8806055
	0.8	0.4504126	0.5066373	0.5712307	0.6487214	0.7507925	0.8882667
	0.9	0.4769375	0.5313945	0.5935381	0.6676228	0.7646303	0.8946201

8.3. Conclusion

In this chapter, by considering the general form of the J-shaped distribution of the random variable X with the *pdf* given by (8.1), and the corresponding *cdf* given by (8.2.1), we have computed its percentile points, for different sets of values of the parameters, $\gamma = 0.1, 0.2, 0.3, 0.4, 0.5, 0.6, 0.7, 0.8, 0.9$, when b = 0.2, 0.5, 1, respectively, which are provided in Tables 8.1–8.3. For our computations, we used the Maple software program.

We believe that the findings of this chapter would be very useful for the practitioners in various fields of studies and further enhancement of research in distribution theory, and its applications.

Chapter 9

Conclusion

9. Introduction

In this handbook, we have studied the J-shaped distributions and their applications. As a motivation, we have discussed several real-world examples which can be modeled through J-shaped distribution. We have presented the mathematical formulation of the family of J-shaped probability distributions which was first proposed by Topp and Leone (1955). We have discussed several variations of Topp–Leone's family of J-shaped distribution.

We have considered the general form of J-shaped distribution and derived its moments, without loss of generality. We also have also investigated other distributional properties of the J-shaped distribution. To describe the shapes of the J-shaped distribution, the plots of the *cdf* and *pdf* for various values of the parameter have been provided. The effects of the parameters such as the J-shape of the *pdf* can easily be seen from these graphs. By taking different values of the scale parameter of the J-shaped distribution, the plots of the moment and variance are presented. Entropy provides an excellent tool to quantify the amount of information (or uncertainty) contained in a random observation regarding its parent distribution (population). A large value of entropy implies the greater uncertainty in the data. As such, Shannon entropy of the J-shaped distribution is provided.

Some distributional properties of order statistics of the J-shaped distribution such as moment, variance, product moments, covariance, and some numerical computations of these for selected values of the parameters are provided. The distributional properties of the record values of the J-shaped distribution are also investigated. Some discussions on the sum, product and ratio of the J-shaped distributions are provided.

The characterizations of the J-shaped distribution are given by using the method of truncated moment, order statistics and record values, where we have considered a product of reverse hazard rate and another function of the truncated point. The derivation of the J-shaped distribution is also given as a solution of the generalized Pearson system of differential equation.

Finally, we have provided the percentile points of the J-shaped distribution for various values of the parameter.

9.1. Conclusion

We believe that the findings of this handbook would be very useful for the practitioners in various fields of studies and further enhancement of research in distribution theory, and its applications.

Appendix

A. Formulas and Results

The following definitions and mathematical results are useful in the study of the J-shaped distributions. For details on these, see, for example, Erdélyi (1953), Abramowitz and Stegun (1972), Prudnikov et al. (1986), Gradshteyn and Ryzhik (2000), and Oldham et al. (2009), among others

A.1. Gamma Function

Definition: The integral given by

$$\Gamma(\alpha) = \int_0^\infty t^{\alpha-1} e^{-t}\, dt, \ (\Re(\alpha) > 0),$$

is called a (complete) gamma function.

The integrals given by

$$\gamma(\alpha, x) = \int_0^x t^{\alpha-1} e^{-t}\, dt, \ (\Re(\alpha) > 0 \text{ and } x \geq 0),$$

and

$$\Gamma(\alpha, x) = \int_x^\infty t^{\alpha-1} e^{-t}\, dt, \ (x \geq 0;\ \Re(\alpha) > 0 \text{ when } x = 0),$$

are called incomplete gamma and complementary incomplete gamma functions, respectively.

Note the following decomposition formula:

$$\Gamma(\alpha, x) + \gamma(\alpha, x) = \Gamma(\alpha), \ (\Re(\alpha) > 0).$$

A.2. Digamma Function

Def. A digamma function, denoted by $\psi(z)$ and also called a psi function, is defined as

$$\psi(z) = \frac{d}{dz}\left[\ln \Gamma(z)\right] = \frac{\Gamma'(z)}{\Gamma(z)}.$$

A.3. Error Function

Definitions.

(ia) The error function, denoted by $erf(x)$, is defined as

$$erf(x) = \frac{2}{\sqrt{\pi}} \int_0^x e^{-u^2} du = P\{|Y| \le x\},$$

where $Y \sim N\left(\mu = 0, \sigma^2 = \frac{1}{2}\right)$ represents the normal distribution with mean $\mu = 0$ and variance $\sigma^2 = \frac{1}{2}$

(ib) The function defined by

$$erfc(x) = \frac{2}{\sqrt{\pi}} \int_x^\infty e^{-u^2} du = 1 - erf(x),$$

is called the complementary error function.

(i) $$\int_{-\infty}^{+\infty} x^2 e^{-x^2/2} dx = \sqrt{2\pi}.$$

(ii) $$\frac{1}{\sqrt{2\pi}} \int_{-\infty}^{0} e^{-u^2/2} du = \frac{1}{2}.$$

(iii) $\frac{d}{dx}\left[erf\left\{\frac{x}{\sqrt{2}}\right\}\right]=\frac{\sqrt{2}}{\pi}e^{-x^2/2}.$

(iv) $\left.\begin{array}{l} x=-\infty \Rightarrow erf(-\infty)=-erf(\infty)=-1, \\ x=+\infty \Rightarrow erf(+\infty)=1. \end{array}\right\}.$

(v) $\gamma\left(\frac{1}{2}, x\right)=\sqrt{\pi}\, erf\left(\sqrt{x}\right).$

(vi) $\Gamma\left(\frac{1}{2}, x\right)=\sqrt{\pi}\, erfc\left(\sqrt{x}\right).$

A.4. Properties of Gamma Function

a. Let $\Gamma(\alpha)$ be a gamma function as defined in (1). Then

(i) $\Gamma(1) = 1$.
(ii) $\Gamma(\alpha) = (\alpha - 1)\,\Gamma(\alpha - 1)$.
(iii) $\Gamma(n) = (n - 1)!$, if n is a positive integer.
(iv) $\Gamma(1/2) = \sqrt{\pi}$
(v) (a) $\Gamma\left[n+\frac{1}{2}\right]=\frac{\pi^{1/2}(2n)!}{2^{2n}(n)!}, n = 0, 1, 2, 3, \ldots$.

b. For negative values, gamma function can be defined as

$$\Gamma\left(-n+\frac{1}{2}\right)=\frac{(-1)^n\, 2^n \sqrt{\pi}}{1.3.5.\ldots(2n-1)},\ \text{where } n\geq 0 \text{ is an integer.}$$

(vi) (vi) $\int_0^\infty t^{\alpha-1}e^{-\frac{t}{\lambda}}dt=\lambda^\alpha.\Gamma(\alpha), \quad \alpha>0.$

c. Let $\psi(z)$ be a digamma function. Then $\psi(z)$ satisfies the following properties:

(i) $\psi(z + 1) = \psi(z) + (1/z)$.
(ii) $\psi(1 - z) = \psi(z) + \pi\cot(\pi z)$.

(iii) $\psi(n) = -\gamma + \sum_{k=1}^{n-1}(k^{-1}), \ \forall$ integer $n \geq 2$,

where

$$\gamma = \lim_{j\to\infty}\left[\{(1 + 1/2 + 1/3 + \ldots + 1/(j-1)\} - ln(j-1)\right] \approx 0.57721566$$

is called the Euler's constant.

(iv) $\psi(1) = -\gamma$.
(v) $\psi(2) = -\gamma + 1$.
(vi) $\psi(1/2) = -\gamma - 2\ln 2$.
(vii) $1/(2z) < \ln(z) - \psi(z) < 1/z, (z > 0)$.
(viii) $1/z < \psi(z) < 1/(z-1), (z > 1)$.
(ix) The function given by

$$\beta(z) = \frac{1}{2}\left[\psi\left(\frac{z+1}{2}\right) - \psi\left(\frac{z}{2}\right)\right],$$

Is called the $\beta(z)$ function.

(x) $\int_0^\infty t^{j-1} e^{-t} ln(t) dt = \Gamma(j).\psi(j), \quad j \geq 1$ is an integer.

(xi) $\int_0^\infty t^{j-1} e^{-t/\theta} ln(t) dt = \theta^j \Gamma(j)[\psi(j) + ln(\theta)], \quad j \geq 1$ is an integer, and $\theta > 0$.

A.5. Additional Formulas

(i) *Shannon Entropy:* For an absolutely continuous random variable X having the probability density function $\varphi_X(x)$, it is defined as

$$H[X] = E\left[-\ln\{\varphi_X(x)\}\right] = -\int_S \varphi_X(x)\ln\{\varphi_X(x)\}\,dx\,,$$

where $S = \{x:\ \varphi_X(x) > 0\}$.

(ii) *Percentile or the Quantile of Order* α**:** For any α, where $0 < \alpha < 1$, the $(100\alpha)th$ percentile or the quantile of order α of a probability distribution with the *pdf* $f_X(x)$ is a number t_α such that the area under $f_X(x)$ to the left of t_α is α. That is, t_α is any root of the equation

$$F(t_\eta) = \int_{-\infty}^{t_\alpha} f_X(u)\,du = \alpha\cdot$$

(iii) *An Useful Binomial Series* (cf. Gradshteyn and Ryzhik (2000), Equation 1.114.1, Page 21)**:**

$$(1+\sqrt{1+u})^m = \left(2^m\right)\left\{1 + \frac{m}{1!}\frac{u}{(4)} + \frac{m\,(m-3)}{2!}\left(\tfrac{u}{4}\right)^2 + \frac{m\,(m-4)(m-5)}{3!}\left(\tfrac{u}{4}\right)^3 + \frac{m\,(m-5)(m-6)(m-7)}{4!}\left(\tfrac{u}{4}\right)^4 + \cdots\right\}$$

where $u^2 < 1$ and m is a real number.

(iv) *An Useful Formula* (cf. Gradshteyn and Ryzhik (2000), Equation 4.331.1, Page 573):

$$\int_0^\infty e^{-\mu x}\ln(x)\,dx = -\frac{1}{\mu}\left[\psi(1) - \ln(\mu)\right],\ \left[\operatorname{Re}\mu > 0\right].$$

(v) *An Useful Formula* (cf. Gradshteyn and Ryzhik (2000), Equation 8.253.1, Page 931):

A series representation of the error function $\Phi(.)$:is given by

$$\Phi(x)=\left(\frac{2}{\sqrt{\pi}}\right)e^{-x^2}\sum_{k=0}^{\infty}\frac{2^k x^{2k+1}}{(2k+1)!!}.$$

(vi) *An Useful Formula* (cf. Gradshteyn and Ryzhik (2000), Equation 4.352.1, Page 576):

$$\int_0^{\infty} x^{\nu-1}e^{-\mu x}\ln(x)\,dx=\frac{1}{\mu^{\nu}}\Gamma(\nu)\left[\psi(\nu)-\ln(\mu)\right],$$

$$\left[\operatorname{Re}\mu>0, \operatorname{Re}\nu>0\right].$$

(vii) *An Useful Formula* (cf. Gradshteyn and Ryzhik (2000), Equation 6.283.2, Page 649) ;

For $\operatorname{Re}(p)>0, \operatorname{Re}(q+p)>0$,

$$\int_0^{\infty}\Phi\left(\sqrt{qt}\right)\,e^{-pt}dt=\frac{\sqrt{q}}{p}\frac{1}{\sqrt{p+q}},$$

where $\Phi(.)$ denotes the error function.

(viii) *An Useful Formula* (cf. Gradshteyn and Ryzhik (2000), Equation 3.381.1, Page 317) :

For $\nu>0$,

$$\int_0^{u} x^{\nu-1}e^{-\mu x}dx=\mu^{-\nu}\,\gamma(\nu,\mu u),$$

where $\gamma(.)$ denotes the incomplete gamma function.

(ix) *An Useful Formula* (cf. Gradshteyn and Ryzhik (2000), Equation 8.352.1, Page 940):

For $n = 0, 1, \ldots$

$$\gamma(1+n, z) = (n!)\left[1 - e^{-z}\left(\sum_{m=0}^{n} \frac{z^m}{m!}\right)\right].$$

References

Abramowitz, M., and Stegun, I. A. (1972). *Handbook of Mathematical Functions, with Formulas, Graphs, and Mathematical Tables.* Dover, New York, USA.

Adamska, J, Bielak, Ł, Janczura, J, Wyłomańska, A. (2022). From Multi- to Univariate: A Product Random Variable with an Application to Electricity Market Transactions: Pareto and Student's t-Distribution Case. *Mathematics*, 10(18):3371. https://doi.org/10.3390/math10183371.

Ahmadi, J., Doostparast, M., and Parsian, A. (2005). Estimation and prediction in a two-parameter exponential distribution based on k-record values under LINEX loss function. *Communications in Statistics–Theory and Methods*, 34(4), 795-805.

Ahsanullah, M. (1978). A characterization of the exponential distribution by spacings. *Journal of Applied Probability*, 15(3), 650-653.

Ahsanullah, M. (1988). *Introduction to Record Statistics*, Ginn Press, Needham Heights, MA, USA.

Ahsanullah, M. (1994). Records of Univariate Distributions. *Pak. J. Statist.* 9(3), 49-72.

Ahsanullah, M. (1995). *Record Statistics.* Nova Science Publishers Inc., New York, NY., USA

Ahsanullah, M. (2004). *Record Values-Theory and Applications.* University Press of America, Lanham, MD, USA.

Ahsanullah, M. (2006). The generalized order statistics from exponential distribution. *Pak. J. Statist.*, Vol. 22, 2, 121-128.

Ahsanullah, M. (2017). *Characterizations of univariate continuous distributions.* Atlantis Press, Paris, France.

Ahsanullah, M., and Aliev, F. (2008). Some characterizations of exponential distribution by record values. *Journal of Statistical Research*, Vol. 2, No.1, 11-16.

Ahsanullah, M., and Nevzorov, V. B. (2001). *Ordered Random Variables.* Nova Publishers, New York, USA.

Ahsanullah, M., and Nevzorov, V. B. (2005). *Order Statistics: Examples and Exercises.* Nova Publishers, New York, USA.

Ahsanullah, M., Nevzorov, V. B (2015). *Records via Probability Theory.* Atlantis Press, Paris, France.

Ahsanullah, M., Hamedani, G. G., and Shakil, M. (2010). On record values of univariate exponential distribution. *Journal of Statistical Research*, 44(2), 267.

Ahsanullah, M., Nevzorov, V. B., and Shakil, M. (2013). *An Introduction to Order Statistics.* Atlantis Press, Paris, France.

Ahsanullah, M., Kibria, B. M. G. and Shakil, M. (2014). *Normal and Student´s t Distributions and Their Applications.* Atlantis Press, Paris, France.

Akbari, M. (2020). Characterization and Goodness-of-Fit Test of Pareto and Some Related Distributions Based on Near-Order Statistics, *Journal of Probability and Statistics,* Article ID 4262574, 1–9.

Al-Hussaini, E. K., and Ahmad, A. E. B. A. (2003). On Bayesian interval prediction of future records. *Test*, 12(1), 79-99.

Ali, M. M., and Nadarajah, S. (2004). On the product and the ratio of t and logistic random variables. *Calcutta Statistical Association Bulletin*, 55, 1 - 14.

Ali, M. M., and Nadarajah, S. (2005). On the product and ratio of *t* and Laplace random variables. *Pakistan Journal of Statistics*, 21(1), 1 - 14.

Ali, M. M., Pal, M. and Woo, J. (2007). On the Ratio of Inverted Gamma Variates. *Aust. J. Statist.,* 36(2), 153-159.

Al-Zahrani, B. (2012). Goodness-of-Fit for the Topp-Leone Distribution with Unknown Parameters. *Applied Mathematical Sciences*, 6(128), 6355-6363.

Amari, S. V., and Misra, R. B. (1997). Closed-form expressions for distribution of sum of exponential random variables. *IEEE Transactions on Reliability*, Vol. 46, 519-522.

Arnold, B. C. and Balakrishnan, N. (1989). Relations, Bounds and Approximations for Order Statistics. *Lecture Notes in Statistics*, No. 53, Springer-Verlag, New York, USA.

Arnold, B. C., Balakrishnan, N. and Nagaraja, H. N. (2005). *A First Course in Order Statistics*. Wiley, New York, USA.

Arnold, B. C., Balakrishnan, N. and Nagaraja, H. N. (1998). *Records*. John Wiley& Sons Inc., New York, USA.

Awad, A. M., and Raqab, M. Z. (2000). Prediction intervals for the future record values from exponential distribution: comparative study. *Journal of Statistical Computation and Simulation*, 65(1-4), 325-340.

Balakrishnan, N., and Nevzorov, V. B. (2003). *A Primer on Statistical Distributions*, John Wiley & Sons, Inc., New York, USA.

Balakrishnan, N., Doostparast, M., and Ahmadi, J. (2009). *Reconstruction of past records,* Metrika, 70, 89–109.

Barr, D. R. and Sherrill, E. T. (1999). Mean and Variance of Truncated Normal Distributions. *The American Statistician*, 53, 357-361. https://doi.org/10.1080/00031305.1999.10474490.

Bluman, A. G. (2023). *Elementary Statistics: A Step by Step Approach*, 11th ed., McGraw Hill LLC, New York, USA.

Bouchrika, I. (2023). *Frequency of Leading Mathematics Scientists Per Country in 2023*, https://research.com/careers/best-mathematics-scientists-2023-report.

Bühlmann, H. (1967). Experience rating and credibility. *Astin. Bull.* 4(03), 199–207.

Brody, J. P., Williams, B. A., Wold, B. J., and Quake, S. R. (2002). Significance and statistical errors in the analysis of DNA microarray data. *Proceedings of the National Academy of Sciences of the United States of America*, 99 (20), pp. 12975-12978. doi: 10.1073/pnas.162468199.

Cigizoglu, H. K., and Bayazit, M. (2000). A generalized seasonal model for flow duration curve. *Hydrological Processes*, 14, 1053 - 1067.

David, H. A. and Nagaraja, H. N. (2003). *Order Statistics*, 3rd ed., John Wiley & Sons, New York, USA.

de Brito, C., Rego, L., de Oliveira, W., and Gomes-Silva, F. (2018). Method for Generating Distributions and Classes of Probability Distributions: The Univariate Case. Hacettepe University Bulletin of Natural Sciences and Engineering Series B: *Mathematics and Statistics*, 48 (2). doi: 10.15672/HJMS.2018.619.

Dudewicz, E. J., and Mishra, S. N. (1988). *Modern Mathematical Statistics*. John Wiley & Sons, New York, USA.

Dunsmore, J. R. (1983). The Future Occurrence of Records. *Ann. Inst. Stat. Math.,* 35, 267-277.

Erdélyi, A. (1953). *Higher Transcendental Functions*, Vols. I and II. McGraw-Hill Book Company, Inc., New York, USA.

Forbes, C., Evans, M., Hastings, N., and Peacock, B. (2010). *Statistical Distributions*, 4th Edition, John Wiley & Sons, Inc., New York, USA.

Foster, F. G. and Stuart, A. (1954). Distribution Free Tests in Time Series Based on the Breaking of Records. *J. Roy. Statist. Soc.*, B, 16, 1-22.

Galambos, J., and Kotz, S. (1978). Characterizations of probability distributions. A unified approach with an emphasis on exponential and related models, *Lecture Notes in Mathematics*, 675, Springer, Berlin, Germany.

Galambos, J. and Simonelli, I. (2005). *Products of Random Variables – Applications to Problems of Physics and to Arithmetical Functions*. CRC Press. Boca Raton / Atlanta, USA.

Garg, M. and Agrawal, J. (2007). On ratio of two independent random variables from different families of distributions. *Journal of Rajasthan Academy of Physical Sciences*. 6.

Genç, A. İ. (2012). Moments of order statistics of Topp–Leone distribution. *Statistical Papers*, 53(1), 117-131.

Genç, A. İ. (2013) Estimation of $P(X>Y)$ with Topp–Leone distribution, *Journal of Statistical Computation and Simulation,* 83:2, 326-339, doi: 10.1080/00949655.2011.607821.

Ghitany, M. E., Kotz, S., and Xie, M. (2005). On some reliability measures and their stochastic orderings for the Topp–Leone distribution. *Journal of Applied Statistics*, 32(7), 715-722.

Ghosh, P. (2005). J-Shaped Distribution. *Encyclopedia of Biostatistics*. John Wiley & Sons, New York.

Giles, D. E. (2012). *A Note on Improved Estimation for the Topp-Leone Distribution* (No. 1203). Department of Economics, University of Victoria, Victoria, BC, Canada.

Glick, N. (1978). Breaking Records and Breaking Boards. *Amer. Math. Monthly*, 85(1), 2-26.

Glänzel, W. (1987). A characterization theorem based on truncated moments and its application to some distribution families, *Mathematical Statistics and Probability Theory* (Bad Tatzmannsdorf, 1986), Vol. B, 75-84, Reidel, Dordrecht, Germany.

Glänzel, W. (1990) Some consequences of a characterization theorem based on truncated moments, *Statistics*, 21,613-618.

Glänzel, W., Telcs, A. and Schubert, A. (1984). *Characterization by truncated moments and its application to Pearson-type distributions*, Z. Wahrsch. Verw. Gebiete, 66, 173-183.

Gradshteyn, I. S., and Ryzhik, I. M. (2000). *Table of Integrals, Series, and Products* (6th Printing). Academic Press, San Diego.

Grubel, H. G. (1968). Internationally diversified portfolios: welfare gains capital flows. *American Economic Review*, 58, 1299 - 1314.

Gulati, S., and Padgett, W. J. (2003). Parametric and Nonparametric Inference from Record Breaking Data. *Lecture Notes in Statistics*, Vol. 172, Springer, New York, NY, USA.

Gunduz, S., and Genc, A. I. (2015). The distribution of the quotient of two triangularly distributed random variables, *Stat Papers*, 56:291–310. doi 10.1007/s00362-014-0582-x.

Hassan, N. J., Nasar, A. H., and Hadad, J. M. (2020). Distributions of the Ratio and Product of Two Independent Weibull and Lindley Random Variables. *Journal of Probability and Statistics,* Volume 2020, 1-8. https://doi.org/10.1155/2020/5693129.

Hayes, R. M. (2004). *J-Shaped Distributions*. http://polaris.gseis.ucla.edu/rhayes/courses/Other/J-Shaped%20Distributions.ppt.

Houchens, R. L. (1984). *Record value theory and inference*. Ann Arbor: University Microfilms International.

Hu, N., Pavlou, P. A., and Zhang, J. (2007). Why do online product reviews have a J-shaped distribution? Overcoming biases in online word-of-mouth communication. *Marketing Science*, 198, 7.

Johnson, N. L., Kotz, L., and Balakrishnian, N. (1994). *Continuous Univariate Distributions*, Volume 1 (Second Edition). John Wiley & Sons, New York, USA.

Johnson, N. L., Kotz, L., and Balakrishnan, N. (1995). *Continuous Univariate Distributions*, Volume 2 (Second Edition). John Wiley & Sons, New York, USA.

Kagan, A. M., Linnik, Yu. L., and C. R. Rao, C. R. (1973). *Characterizations Problems in Mathematical Statistics*, John Wiley, New York, USA.

Kamps, U. (1995). A concept of generalized order statistics. *Journal of Statistical Planning and Inference*, 48(1), 1-23.

Kapadia, A. S., Chan, W., and Moyé, L. A. (2005). *Mathematical Statistics with Applications*. Chapman & Hall/CRC, Boca Raton, USA.

Khoolenjani, N. B., and Khorshidian, K. (2009). On the Ratio of Rice Random Variables. *Journal of the Iranian Statistical Society*, 8, 61-71.

Kibria, B. G., and Nadarajah, S. (2007). Reliability Modeling: Linear Combination and Ratio of Exponential and Rayleigh. *IEEE Transactions on Reliability*, 56(1), 102-105.

Kim, J. H., and Jeon, Y. (2013). Credibility theory based on trimming. *Insur. Math. Econ.* 53(1), 36–47.

Klimczak, M., and Rychlik, T. (2005). Reconstruction of previous failure times and records. *Metrika,* 61, 277–290.

Kordonsky, K. B., and Gertsbakh, I. B. (1995). System state monitoring and lifetime scales-I, *Reliability Engineering & System Safety*, vol. 47, pp. 1–14.

Kordonsky, K. B., and Gertsbakh, I. B. (1995). System state monitoring and lifetime scales-II, *Reliability Engineering & System Safety*, vol. 49, pp. 145–154.

Korhonen, P. J. and Narula, S. C. (1989). The probability distribution of the ratio of the absolute values of two normal variables. *Journal of Statistical Computation and Simulation,* 33, 173–182.

Kotz, S., and Nadarajah, S. (2006) J-shaped distribution, Topp and Leone's. *Encyclopedia of statistical sciences,* vol 6, 2nd edn. Wiley, New York, p 3786.

Kotz, S., and Seier, E. (2007). *Kurtosis of the Topp-Leone distributions*. InterStat, July, 2007.

Kotz, S., and Shanbhag, D. N. (1980). Some new approaches to probability distributions. *Advances in Applied Probability*, 12, 903-921.

Kotz, S., and van Dorp, J. R. (2004). Uneven two-sided power distributions with applications in econometric models. *Statistical Methods and Applications*, 13(3), 285-313.

Koudou, A. E., and Ley, C. (2014). *Characterizations of GIG laws: A survey, Probability surveys,* 11, 161 – 176.

Kumar, D. (2014). Explicit Expressions for Moments of K−TH Lower Record Values from J-Shaped Distribution and a Characterization. *Mathematical Sciences Letters*, 3, No. 3, 237-241.

Ladekarl, M., Jensen, V., and Nielsen, B. (1997). Total number of cancer cell nuclei and mitoses in breast tumors estimated by the optical disector. *Analytical and Quantitative Cytology and Histology*, 19, 329 - 337.

Larson, R. J., and Marx, M. L. (2006). *An Introduction to Mathematical statistics and its Applications*. Pearson Prentice Hall, New Jersey, USA.

Leak, W. B. (1965). The J-shaped probability distribution. *Forest Science*, 11(4), 405-409.

Leiserson, G., McGrew, W., and Kopparam, R. (2019). *Net worth taxes: What they are and how they work*, Washington Center for Equitable Growth, Washington, USA. (https://equitablegrowth.org/the-distribution-of-wealth-in-the-united-states-and-implications-for-a-net-worth-tax/).

Lee, R. Y., Holland, B. S. and Flueck, J. A. (1979). Distribution of a ratio of correlated gamma random variables. *SIAM Journal on Applied Mathematics*, 36, 304–320.

Lukacs, E. (1972). *Probability and Mathematical Statistical Statistics – An Introduction*. Academic Press, New York, USA.

Mallick, A., Ghosh, I., and Hamedani, G. G. (2018). A note on Sum, Difference, Product and Ratio of Kumaraswamy Random Variables. *Journal of Statistical Theory and Applications*, Volume 17, Issue 2, Pages 230 – 241.

Marsaglia, G. (1965). Ratios of normal variables and ratios of sums of uniform variables. *Journal of the American Statistical Association*, 60, 193–204.

McGraw-Hill *Dictionary of Scientific and Technical Terms* (2003), 6th ed, McGraw-Hill, New York, USA.

Miloševic, B. (2017). Some recent characterization-based goodness of fit tests. In *the Proceedings of the 20th European Young Statisticians Meeting*, (pp. 67-73), (Uppsala, Sweden).

Nadarajah, S. (2005). Linear combination, product and ratio of normal and logistic random variables. *Kybernetika,* Vol. 41, No. 6, 787 – 798.

Nadarajah, S. (2005a). Sums, Products, and Ratios of Non-central Beta Variables. *Communications in Statistics - Theory and Methods*, 34:1, 89-100, doi: 10.1081/STA-200045865.

Nadarajah, S. (2005b). Products, and ratios for a bivariate gamma distribution. *Applied Mathematics and Computation*, 171(1), 581 – 595.

Nadarajah, S. (2005c). On the Product and Ratio of Laplace and Bessel Random Variables. *Journal of Applied Mathematics*, 4, 393 - 402.

Nadarajah, S., and Ali, M. M. (2004). On the product of Laplace and Bessel random variables. *Journal of the Korean Data & Information Science Society*, 15(4), 1011 – 1017.

Nadarajah, S., and Ali, M. M. (2005). On the Product and Ratio of t and Laplace Random Variables. *Pakistan Journal of Statistics*, 21, 1 – 1.

Nadarajah, S., and Gupta, A. A. (2005). On the Product and Ratio of Bessel Random Variables. *International Journal of Mathematics and Mathematical Sciences*, 18, 2977 – 2989.

Nadarajah, S., and Gupta, A. A. (2006), On the ratio of logistic random variables. *Computational Statistics & Data Analysis*, 50, 1206 – 1219.

Nadarajah, S., and Kotz, S. (2005). On the Product and Ratio of Pearson Type VII and Laplace Random Variables. *Austrian Journal of Statistics*, Vol. 34 (1.1), 11– 23.

Nadarajah, S., and Kotz, S. (2005). On the Linear Combination of Exponential and Gamma Random Variables. *Entropy*, 7(2), 161 – 171.

Nadarajah, S., and Kotz, S. (2006). The Linear Combination of Logistic and Gumbel Random Variables. *Fundam. Inform*, 74(2-3), 341 – 350.

Nadarajah, S., and Kotz, S. (2006), On the Ratio of Fr´echet Random Variables. *Quality & Quantity*, 40,861–868.

Nadarajah, S., and Kotz, S. (2006). On the product and ratio of gamma and Weibull random variables. *Econometric Theory*, 22, 338 – 344.

Nadarajah, S., and Kotz, S. (2007). Generalized financial ratios. *Mathematical Methods in the Applied Sciences*, 30.

Nadarajah, S. (2008). A Review of Results on Sums of Random Variables. *Acta Appl Math* 103, 131–140. https://doi.org/10.1007/s10440-008-9224-4.

Nadarajah, S. (2009). Bathtub-shaped failure rate functions. *Quality and Quantity*, 43, 855-863.

Nadarajah, S. (2010). Distribution properties and estimation of the ratio of independent Weibull random variables. *AStA Adv Stat Anal,* 94, 231–246. https://doi.org/10.1007/s10182-010-0134-1.

Nadarajah, S., and Kotz, S. (2003). Moments of some J-shaped distributions, *Journal of Applied Statistics,* 30,311-317.

Nadarajah, S., and Kotz, S. (2011). On the linear combination, product and ratio of normal and Laplace random variables. *Journal of the Franklin Institute*, Volume 348, Issue 4, Pages 810-822. https://doi.org/10.1016/j.jfranklin.2011.01.005.

Nagar, D. K., and Ramirez-Vanegas, Y. A. (2013). Distributions of Sum, Difference, Product and Quotient of Independent Non-central Beta Type 3 Variables, *British Journal of Mathematics & Computer Science*, 3(1): 12-23.

Nagaraja, H. (2006). Characterizations of Probability Distributions, In Springer *Handbook of Engineering Statistics*, Springer, London, UK, 79 - 95.

Nagaraja, H. N. (1988) Record Values and Related Statistics -- a Review, *Commun. Statist. Theory-Methods*, 17, 2223-2238.

Nason, G. (2006). On the sum of t and Gaussian random variables. *Statistics & Probability Letters*, 76, 1280 – 1286.

Nevzorov, V. B. (1988). Records. *Theo. Prob. Appl.,* 32, 201-228.

Nevzorov, V. B. (2001). Records: Mathematical Theory. *Translation of Mathematical Monographs*, Volume 194. American Mathematical Society. Providence, RI, USA.

Nikitin, Y. Y. (2017). *Tests based on characterizations, and their efficiencies: A survey*, Acta et Commentationes Universitatis Tartuensis de Mathematica, 21(1), 3-24.

Obeid, N., and Kadry, S. (2019). On The Product and Ratio of Pareto and Rayleigh Random Variables. *Pakistan Journal of Statistics*, 35(4), 285-300.

Obeid, N., and Kadry, S. (2021a). Distributions of the Product and Ratio of Two Independent Pareto and Beta Random Variables. *J. Stat. Appl. Pro.*, Vol. 10, No. 2, PP:291-304. doi:10.18576/jsap/100204.

Obeid, N., and Kadry, S. (2021b). *A Review of Results on Ratios of Different Families Random Variables.* Preprint, February 2021. doi:10.13140/RG.2.2.15082.44480.

Obeid, N. and Kadry, S. (2022). Distributions of the Product and Ratio of Two Independent Pareto and Exponential Random Variables. *J. Stat. Appl. Pro.*, 11, No. 1, 291-304. http://dx.doi.org/10.18576/jsap/110123.

Oldham, K. B., Myland, J., and Spanier, J. (2009). *An Atlas of Functions with Equator, the Atlas Function Calculator.* Springer, New York, USA.

Patel, J. K., Kapadia, C. H., and Owen, D. B. (1976). *Handbook of Statistical Distributions*, Marcel Dekker, New York, USA.

Pham-Gia, T. (2000). Distributions of the ratios of independent beta variables and applications. *Communications in Statistics—Theory and Methods*, 29, 2693–2715.

Press, S. J. (1969). The t ratio distribution. *Journal of the American Statistical Association*, 64, 242–252.

Prudnikov, A. P., Brychkov, Y. A., and Marichev, O. I. (1986). *Integrals and Series*, Volumes 1, 2, and 3, Gordon and Breach Science Publishers, Amsterdam, Netherlands.

Rao, C. R., and Shanbhag, D. N. (1998). Recent Approaches for Characterizations Based on Order Statistics and Record Values. *Handbook of Statistics*, (N. Balakrishnan and C. R. Rao eds.) Vol. 10, 231-257.

Renyi, A. (1962). Theorie des Elements Saillants d'une Suit d'observations Collq. Combinatorial Meth. *Prob. Theory,* Aarhus University, 104-115.

Resnick, S. I. (1973). Record Values and Maxima. *Ann. Probab.*, 1, 650-662.

Rohatgi, V. K., and Saleh, A. K. M. E. (2001). *An Introduction to Probability and Statistics*. John Wiley & Sons, New York, USA.

Rokeach, M., and Kliejunas, P. (1972). Behavior as a function of attitude-toward-object and attitude-toward-situation. *Journal of Personality and Social Psychology*, 22, 194 - 201.

Samaniego, F .J. and Whitaker, L. D. (1986). On Estimating Population Characteristics from Record Breaking Observations. I. Parametric *Results. Naval Res. Log. Quart.*, 25, 531- 543.

Shakil, M., Kibria, B. M. G., and Singh, J. N. (2006). Distribution of the ratio of Maxwell and Rice random variables. *Int. J. Contemp. Math. Sci.,* Vol. 1, 2006, No. 13, 623 – 637.

Shakil, M., and Kibria, B. M. G. (2006). Exact Distribution of the Ratio of Gamma and Rayleigh Random Variables. *Pakistan Journal of Statistics and Operation Research*, 2(2), 87-98. https://doi.org/10.18187/pjsor.v2i2.91.

Shakil, M., and Kibria, B. M. G. (2007). On the product of Maxwell and Rice random variables. *Journal of Modern Applied Statistical Methods*, Volume 6, No.1, 212-218.

Shakil, M., Kibria, B. M. G., and Chang, K. C. (2008). Distributions of the product and ratio of Maxwell and Rayleigh random variables. *Statistical Papers,* 49(4), 729 - 747.

Shakil, M., and Kibria, B. M. G. (2009). Exact Distributions of the Linear Combination of Gamma and Rayleigh Random Variables. *Austrian Journal of Statistics*, Volume 38, Number 1, 33–44.

Shakil, M., and Ahsanullah, M. (2011). Record values of the ratio of Rayleigh random variables. *Pakistan Journal of Statistics*, 27, 307-325.

Shakil, M., Ahsanullah, M., and Kibria, B. M. G. (2018). On the Characterizations of Chen's Two-Parameter Exponential Power Life-Testing Distribution, *Journal of Statistical Theory and Applications,* 17 (3), 393 - 407.

Shakil, M., Khadim, A., Saghir, A., Ahsanullah, M., and Kibria, B. M. G., and Bhatti, M. I. (2022). Ratio of Two Independent Lindley Random Variables. *J Stat Theory Appl,* 21, 217–241. https://doi.org/10.1007/s44199-022-00050-4.

Shannon, C. E. (1948). A mathematical theory of communication. *Bell System Tech. J.,* Vol. 27, 379 – 423; 623 – 656.

Shorrock, S. (1973). Record Values and Inter-Record Times. *J. Appl. Prob.,* 10, 543-555.

Sindhu, T. N., Saleem, M., and Aslam, M. (2013). Bayesian Estimation for Topp-Leone Distribution under Trimmed Samples, *Journal of Basic and Applied Scientific Research,* 3(10), 347-360.

Sornette, D. (1998). Multiplicative processes and power laws. *Physical Review E*, 57, 4811 – 4813.

Springer, M. D. (1979). *The Algebra of Random Variables.* John Wiley & Sons, New York, USA.

Tahir, M. H., and Nadarajah, S. (2015). Parameter induction in continuous univariate distributions: Well-established G families. *Anais da Academia Brasileira de Ciência*s, 87 (2), 539 - 568. doi: 10.1590/0001-3765201520140299.

Tahir, M., and Cordeiro, G. M. (2016). Compounding of distributions: a survey and new generalized classes. *Journal of Statistical Distributions and Applications,* 3 (13). doi: 10.1186/s40488-016-0052-1.

Thornton, A., Morley, C., Green, S., Cole, T., Walker, K., and Bonnett, J. (1991). Field trials of the Baby Check score card: mothers scoring their babies at home, *Archives of Disease in Childhood*, 66, 106-110.

Tibrewal, S., and Foss, M. (1991). Day care surgery for the correction of hallux valgus, *Health Trends*, Vol 23, No. 3, 117-119.

Topp, C. W. and Leone, F. C. (1955). A Family of J-shaped frequency functions, *Journal of the American Statistical Association,* 50, 209-219.

Volkova, K. Y., and Nikitin, Y. Y. (2015). Exponentiality tests based on Ahsanullah's characterization and their efficiency, *Journal of Mathematical Sciences*, 204(1), 42-54.

Yule, G. U., and Kendall, M. G. (1950). *An Introduction to the Theory of Statistics*, 14th Edition (5th Impression, 1968), Griffin, London, UK.

Zghoul, A. A. (2010). *Order statistics from a family of J-shaped distributions*. METRON, 68(2), 127-136.

Zghoul, A. A. (2011). Record values from a family of J-shaped distributions. *Statistica,* 71(3), 355-365.

Zhou, M., Yang, D. W., Wang, Y., and Nadarajah, S. (2006). Some j-shaped distributions: sums, products and ratios. In Reliability and Maintainability Symposium, 2006. *RAMS'06. Annual* (pp. 175-181). IEEE. http://en.wikipedia.org/wiki/Shape_of_the_distribution. https://epilab.ich.ucl.ac.uk/coursematerial/statistics/index.html. http://privatewww.essex.ac.uk/~scholp/distrib.htm.

Index

U

V

About the Authors

Dr. Mohammad Ahsanullah, PhD, Professor Emeritus

Rider University, Lawrenceville, NJ, USA

Dr. Mohammad Ahsanullah is a Professor Emeritus of Rider University. He earned his PhD from North Carolina State University, Raleigh, North Carolina. He is a Fellow and life member of American Statistical Association and fellow of Royal Statistical Society. He is an elected member of the International Statistical Institute and a life member of the Institute of Mathematical Statistics.. He was feuding member and editor-in- chief of *Journal of Applied Statistical Science and Journal of Statistical Theory and Applications*. He has authored and co-authored more than 50 books and published more than 400 research articles in reputable journals. His research areas are Record Values, Order Statistics, Statistical Inferences, Characterizations of Distributions etc..

Dr. Mohammad Shakil, PhD, Professor

Department of Mathematics,
Miami Dade College, Hialeah, FL, USA

Dr. M. Shakil, PhD, CStat (Royal Statistical Society, UK), PStat (American Statistical Association, USA), is a Professor of Mathematics and Statistics, Miami Dade College, Hialeah Campus, Miami, Florida, USA. He has been teaching mathematics and statistics courses at Miami Dade College, since 2004. Prior to that, he taught mathematics courses at several institutions in USA and overseas. He has co-authored three statistics textbooks and over 130 research articles in reputable journals. His research interests are Distribution Theory, Characterizations, Statistical Inferences, Record Values, Numerical Analysis, etc. He is a member of many international learned societies, and associated with the editorial boards of various journals.